FOOD SCIENCE AND TECHNOLOGY

ADVANCES IN CEREAL AND PSEUDOCEREAL RESEARCH FOR FUNCTIONAL FOODS

FOOD SCIENCE AND TECHNOLOGY

Additional books in this series can be found on Nova's website under the Series tab.

Additional e-books in this series can be found on Nova's website under the e-book tab.

FOOD SCIENCE AND TECHNOLOGY

ADVANCES IN CEREAL AND PSEUDOCEREAL RESEARCH FOR FUNCTIONAL FOODS

NAOFUMI MORITA
PHAM VAN HUNG
AND
TOMOKO MAEDA
EDITORS

nova publishers
New York

For permission to use material from this book please contact us:
Telephone 631-231-7269; Fax 631-231-8175
Web Site: http://www.novapublishers.com

Library of Congress Cataloging-in-Publication Data

Advances in cereal and pseudocereal research for functional foods / editors, Naofumi Morita, Pham Van Hung and Tomoko Maeda (Department of Food Packaging Technology, Toyo College of Food Technology, Hyogo, Japan, and others).
pages cm
Includes index.
ISBN 978-1-62618-347-6 (hardcover)
1. Functional foods. 2. Cereal products. 3. Enriched cereal products. 4. Cereal products industry. I. Morita, Naofumi. II. Van Hung, Pham. III. Maeda, Tomoko.
QP144.F85A38 2013
613.2--dc23

2013007544

Published by Nova Science Publishers, Inc. † New York

CONTENTS

PREFACE

The global economy is in recession presently, especially in the USA and Europe, while developing countries among ASEAN members are achieving economic growth. High quality food materials and ingredients are in great demand from these developing countries and the USA and Europe as well. Cereal grains of corn, wheat and rice are consumed as staple foods. In 2011 the total production of these grains amounted to 21 billion tons worldwide. Global wheat production in the same year amounted to 691 million tons; however, wheat production is not stable. It is often affected by serious climatic problems caused by desertification, devastating flood, drought, etc. Nevertheless most of the people living in industrialized as well as in developing countries wish to eat bread, especially white bread. Milling techniques remove the outer layers of grains to obtain white and delicious flours. As lifestyles change worldwide, however, consumers become more health-conscious and wish to eat more safe and functional foods. Wheat contains important nutrients in the whole grain, as is well known.

From these points of view, the authors developed a polished-grading technique to obtain wheat flour in a stepwise manner from the outer surface of the grain: the obtained flours are different from conventional whole flour. These flours contain more or less germ and epidermis, keeping enough amounts of vitamins, minerals and dietary fiber. With these flours normal baking methods may not show good results, but sourdough bread or partial substitutions with graded flours were found to yield favorable results in the presence of hemicellulase or other ingredients. Wheat flour contains starch (about 75%) as a major component. Flours containing waxy starch (amylose-free) or high-amylose starch have been produced in Japan by genetic recombination. Therefore, recent research on characteristics and functionality of waxy and high-amylose wheats and their application to food processing are described.

Buckwheat, quinoa and amaranth are pseudocereals, and flours obtained from them contain various nutrients, minerals and functional materials. To characterize these functional materials in buckwheat grains, a gradual milling method was applied for whole buckwheat grains from the outer layer in a stepwise fashion. Characteristics of the flour were tested and these flours were applied for noodle making. When these grains were germinated, the functionality of the flours was improved still more and allergenic proteins almost disappeared. Therefore, germinated buckwheat grains were used for processing buckwheat *natto* and *miso* paste.

Quinoa and amaranth are minor crops, but health-conscious people are willing to use them as alternatives to wheat. Therefore, after germination of these grains, functional properties, application to breadmaking, lipid changes and IgE-Immunoblotting reaction were investigated.

Starch is a main component in cereal grains and an important ingredient in daily food. However, this native starch is not suitable for food processing because of retrogradation and gelatinization temperature, etc. Therefore, various kinds of modifications to starches have been made including enzymatic or acid conversion, physical changes such as pre-gelatinization and heat-moisture, and most importantly, chemical treatments including cross-linking, hydroxypropylation and acetylation. These chemically modified starches have been widely used for various prepared foods such as snack foods, bread and cakes to improve their quality. Therefore, the dough properties and breadmaking qualities using hydroxypropylated, acetylated and/or cross-linked corn starches to substitute for commercial wheat flour were reviewed.

Currently the most important aims for food research are to provide consumers food materials that are nutritious, safe, healthy, environmentally friendly, tasty, etc. Therefore, the authors hope that this book is helpful for various food researchers and investigators.

Section 1: Wheat Research and Developments for Functional Foods

In: Advances in Cereal and Pseudocereal ... ISBN: 978-1-62618-347-6
Editors: N. Morita, P.V. Hung et al.

Chapter 1

POLISHED-GRADING TECHNIQUE

***Tomoko Maeda*[1,*] *and Naofumi Morita*[2,†]**
[1]Hyogo University of Teacher Education, Hyogo, Japan
[2]Department of Food Packaging Technology,
Toyo College of Food Technology, Kawanishi, Hyogo, Japan

ABSTRACT

The appearance of all the graded flours was distinctly more brownish than that of a conventionally milled soft-type flour, 'N61', hard-type flour, 'Hermes' and No. 1 Canada Western Red Spring (1CW) flour, 'CW'. The graded flours contained larger amounts of dietary fiber, ash and minerals (K, Ca and Fe) than N61, Hermes or CW. In addition, they showed a higher diastatic activity and contained a larger amount of damaged starch as compared with the common flours. Any kinds of graded flours alone could not make a good gluten structure and the dough was clearly hard and not extensible. The loaves baked with the graded flours alone were smaller and the bread crumbs were harder than those of common flours. The 10 % substitution of soft-type graded flours for N61 improved the dough and baking properties. Moreover, the addition of hemicellulase (HEM), cellulase (CEL) and pentosanase (PEN) to the substituted N61 flours increased the loaf volume and softness of the

* Department of Science, Technology, and Human Life, Graduate School of Education, Hyogo University of Teacher Education, 942-1, Shimokume, Kato, Hyogo 673-1494, Japan. E-mail: mtomoko@hyogo-u.ac.jp.

† Department of Food Packaging Technology, Toyo College of Food Technology, 4-23-2, Mimami-Hanayashiki, Kawanishi 666-0026, Hyogo, Japan.

bread. The combined addition of hard-type graded flours and sucrose fatty acid ester (SE) to CW increased the amount of gas generation and maturity of the dough with the high hydrophilic property. All the graded flours contained larger amounts of water-soluble pentosan (WSP) and water-insoluble pentosan (WISP) than those of common flours. The substitution of the graded flours for N61 increased the amounts of WSP, while the ratio of Ara/Xyl (arabinose to xylose) of WSP decreased. The improvement of graded flours for breadmaking was caused by large amounts of ferulic acid and low Ara/Xyl, which are related to the gel-forming ability of WSP and to the presence of heat stable foam. This stable foam accelerated the extensibility and maturity of the dough, resulting in the production of end products of good quality. In addition, the combination of graded flours and PEN, the presence of enough amount of WSP, and degradation of WSP and WISP having low Ara/Xyl make the dough extensible and viscous, resulting in favorable end products.

1. Structure of Wheat Grain and Preparation of Graded Flours

1.1. Structure and Composition of Wheat Grain

The wheat kernel or grain, known botanically as a caryopsis, is the fruit of the plant. The kernel consists mainly of 3 parts: bran, endosperm and germ [Pomeranz, 1988a]. It is normally about 4-8 mm (in average 6.2 mm) long, 1.4-4.7 mm (2.7 mm) wide, 1.25-1.40 (1.35) specific gravity and approximately 0.03 g weight, depending on the variety and conditions of growth [Nagao, 1996]. The relationship between the botanical constituents and major milling fractions, the color of the kernel is primarily affected by the components present in the pericarp or seed coat [Pomeranz, 1988a]. The pericarp consists of an outer epidermis and hypodermis, next to the layer of thin-walled cells and several other types of cells. Altogether, the pericarp is about 50 μm thick. Then, we find another thin seed coat, covering nucellar epidermis, and then an aleurone layer before coming to the starch-rich endosperm. Bran of the milling fractions is chiefly the outer material down to and including the aleurone layer [Pomeranz, 1988a]. The starch endosperm is the material from which white flour is made. Aleurone layer is also a part of endosperm, but it is separated from bran layers in the milling process. It comprises starch granules (64-74 %) embedded in a matrix of proteins (8.3-

13.0 %) [Pomeranz, 1988a]. The proteins consist of albumins, globulins, gliadins and glutenins. The combination of the gliadins and glutenins is referred to as the gluten complex and is regarded as storage protein [Pomeranz, 1988a]. On the other hand, the wheat germ consists of several parts. The plumule is one part of embryonic axis and forms a shoot when the seed germinates. There is also the scutellum, which is the storage, digestive and absorbing organ, and attached to the plumule. It contains food for the plant, which is supplied at the time of germination, and also transfers further food from the endosperm. The germ is also readily separated from endosperm together with bran by roll milling. But it is an important dietary supplement, providing a rich source of vitamin E [Cornell and Hoveling, 1998]. With regard to various compositions of each milling fraction, bran and germ contain larger amounts of nutrients and vitamins, compared with endosperm [Pomeranz, 1988a].

1.2. Milling Process of Wheat Grain

Concerning a milling method, a normal milling contains processes of grinding and separating steps. Grinding is done on break, sizing and reduction rolls. Separation is carried out using machines called sifters or purifiers. In principle, the milling process may be divided into the seven distinct operating systems, which are break, grading, purification, sizings (scratch), reduction, flour dressing and millfeed systems [Pomeranz, 1988a]. Wheat delivered to a mill usually requires much more extensive cleaning to remove foreign or harmfull materials such as stones, mud, ergot, metals and other seeds, which might adversely affect the appearance or functionality of milled product, and might even damage the mill itself. Hence, before milling, wheat is transferred from the elevator to the wheat cleaning section, often referred to as the "screen room". The cleaned wheat is then prepared for milling by being conditioned (tempered) with an appropriate addition of water [Pomeranz, 1988b]. Conditioning (or tempering) is a controlled addition of moisture to wheat to achieve the following objectives:

1) To harden the pericarp to prevent it from powdering during the milling process (powdered bran cannot be separated from flour at any stage in the milling process);
2) To facilitate the physical separation of endosperm from bran;
3) To mellow the endosperm so that it may be easily to reduce flour;

4) To ensure that all materials leaving the grinding rolls are in optimum condition for sifting; and
5) To ensure that grinding produces the optimum level of damaged starch consistent with the hardness of the wheat and the end use of flour [Pomeranz, 1988b].

The milling process of wheat flour by a conventional method is illustrated simply. At first, about 15% water is added to the wheat grain, and conditioning is done to remove the hard bran from the soft endosperm more easily. Then, breaking, sieving and purification are carried out. In this stage, the prepared flour is very hard with a large particle size and contains some bran.

Therefore, the flour is called ‘semolina’. Moreover, this semolina is gradually ground through smooth rolls in the reduction system to produce the finished flour. Finally, these finished flours are divided depending on the quality of flours, dressed and the commercial wheat flour is prepared. As shown in this procedure, the conventional method can remove the bran and germ from the wheat grain. Therefore, the recovery is approximately 70%.

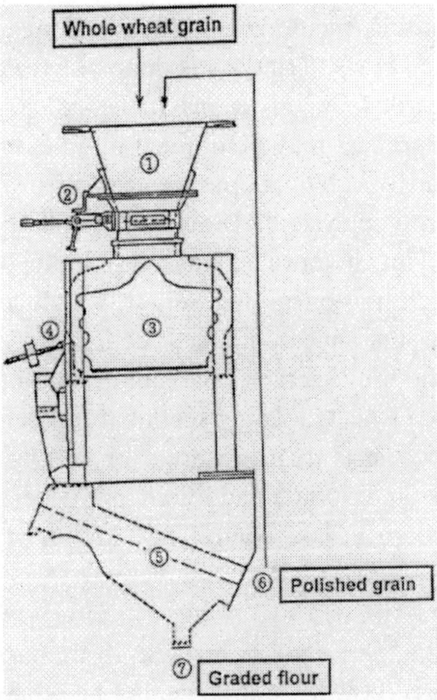

1, Receiver; 2, Lever; 3, Roller; 4, Controller; 5, Sieve; 6, Outlet for polished grain; 7, Outlet for graded flour.

Figure 1-1-1. Milling process of the graded flour by the polished-grading method.

Table 1-1-1. Graded flours prepared from various cultivars

Cultivar					
Norin 61		Norin 61		1CW	
Sample	Fraction (%)	Sample	Fraction (%)	Sample	Fraction (%)
NA-1	100-90	NB-1	100-90	C-1	100-90
NA-2	90-80	NB-2	90-80	C-2	90-80
NA-3	80-70	NB-3	80-70	C-3	80-70
NA-4	70-60	NB-4	70-60	C-4	70-60
NA-5	60-50	NB-5	60-50	C-5	60-50
NA-6	50-40	NB-6	50-40	C-6	50-40
NA-7	40-30	NB-7	40-30	C-7	40-30
NA-8	30-0	NB-8	30-0	C-8	30-0

Hermes, conventionally milled hard-type wheat flour; N61, conventionally milled soft-type wheat flour of Norin 61, CW, conventionally milled hard-type wheat flour of 1CW.

NA and NB are sieved through a pore size of 125 and 600 mm, respectively.

NA and NB are soft-type graded flours obtained from Norin 61and C is hard-type graded flours obtained from 1CW.

NA-1 (100-90%), 10% layer graded from outer part of whole grains; NA-2 (90-80%), 10 % layer graded from outer part of NA-1; NA-3~NA-7, same 10 % grading as above; NA-8 (30-0%), 30%-core of the whole grains. NB-1~NB-8 and C-1~C-8 are prepared by the same grading method as in NA-1~NA-8.

Figure 1-1-1 shows the procedure for a polishing process by the polished-grading method, which we used in this study. This apparatus is a modified Japanese rice-polisher. At first, a whole wheat grain is put into position 1, and the grain is gradually polished from the outer layer in a stepwise manner by the vertical roller located at position 3. Then the polished-graded flours are sieved through the sifter in position 5, and only the passed flour is recovered from position 7 as a graded flour. For example, 100 Kg of wheat grain is put into the receiver (position 1). If 10% of the original weight, namely 10 Kg of flour is recovered from position 7, the flour corresponds to 100 to 90 % of the whole grain. Likewise, the next 10 Kg flour recovered from position 7 corresponds to 90 to 80 % of the whole grain. In this way, the polished-grading process is quite simple and this method can produce the wheat flours from the whole wheat grain with theoretically 100 % of recovery.

Therefore, if we prepare the wheat flour using the polished-grading method, it is no need to discard the bran and germ, which have been removed or used for animal foods until now. Moreover, we can obtain the special wheat flour " the graded flour" derived from the whole wheat grain, containing large amounts of dietary fiber, vitamins and minerals. Accordingly, the graded flours are sure to produce end products with high safety, deliciousness and nutrition, which are essential factors to be needed for foods. These end products must sufficiently satisfy the appetite of our consumers. Therefore, the graded flours seem to be expected as a new foodstuff.

1.3. Preparation of Various Graded Flours by the Polished-Grading Method

Table 1-1-1 shows the various graded flours used for this experiment. For the case of graded flours obtained from Norin 61, the graded flour samples were sieved through a pore size of 125 μm (named as 'NA') or 600 μm (named as 'NB'). For the case of hard-type graded flours, capital letter of 'C' was attached to the name. NA-1, NB-1 and C-1 correspond to the graded flour from 100-90% of the whole grain; NA-2, NB-2 and C-2, 90-80%; NA-3, NB-3 and C-3, 80-70%; NA-4, NB-4 and C-4, 70-60%; NA-5, NB-5 and C-5, 60-50%; NA-6, NB-6 and C-6, 50-40%; NA-7, NB-7 and C-7, 40-30%; NA-8, NB-8 and C-8, 30% to the core. These graded flours were used for the following experiments.

2. Characterization of Soft-Type Polished-Graded Wheat Flours

Characteristics of wheat flour graded by polishing and the effect of the graded flour substitution for conventionally milled flour on the physical properties of wheat dough and bread were studied. When wheat grains are milled by the conventional method, most of the fibers and minerals are removed. For this reason, a polished-grading method which keeps whole part of grains was introduced. Wheat grains were gradually polished in a stepwise manner from the outer layer by 10% of the total weight to the core of the original soft-type wheat grain Norin 61 (N61). The loaf volumes of all graded flours were distinctly smaller than that of conventionally milled N61 and the bread crumbs were quite hard. All graded flours clearly increased viscoelastic properties such as compression stress, modulus of elasticity and viscosity coefficient. The graded flours had a higher SH content, whereas there was a lower SS content in dough than conventionally milled hard-type flour. Any fraction of the polished-graded flours alone did not have good dough and baking properties (Figures 1-1-2, 1-1-3). When 10% of N61 was substituted with a fraction of the graded flour, A-4 (70-60 % of original grains) or A-7 (40-30 %), the specific volume increased slightly as compared with that of N61 alone. Moreover, the addition of hemicellulase (HEM) to the combined flour improved the loaf volume distinctly, and the bread crumbs became quite soft. Microscopic observation of the dough showed that the gluten matrix became thicker and most of the starch granules were sufficiently covered with the matrix, as compared with N61 alone. From these results, a 10% substitution of graded flours A-4 and A-7 to N61 with or without hemicellulase was found to improve some of the viscoelastic and baking properties of wheat dough.

3. Effect of Polished-Graded Soft-Type Wheat Flour Substitution for Commonly Milled Wheat Flour on the Breadmaking

The optimum conditions for breadmaking composed of various amounts of polished-graded flour and N61 were studied. The viscoelastic properties of the doughs were also tested. When 30 or 50% of N61 was substituted with the 70~40 % graded flour (fraction B) of the whole wheat grain, the loaf volume and softness of bread crumbs increased with optimum amounts of water of 75 and 85%, respectively. Moreover, the addition of pentosanase (PEN) or cellulase (CEL) to the substituted N61 distinctly improved the loaf volume and retarded the retrogradation of bread crumbs as compared with those of N61 alone. Especially, the addition of PEN or CEL to the 50%-substituted N61 having 85% water clearly increased viscoelastic parameters, such as the modulus of elasticity (γ) and viscosity coefficient (η), as compared with those of N61 alone. In Farinograph data, 50%-substituted N61 also decreased the arrival and development times but increased the stability time. For the DSC result, 50% substitution tended to decrease gelatinization enthalpy with the addition of PEN or CEL. Microscopic observation of the 50% graded- flour-substituted N61 containing PEN showed that most of the starch granules were covered with extensible gluten and expanded still more during fermentation. In the case of the dough with the optimum water content, the gluten matrix was more fibrous and continuous, followed by sufficient gelatinization of starch granules as compared with N61 alone. Therefore, for the case of substituted N61 flour with graded flour, the optimum amount of water and hydrolysis of pentosan or cellulose in the endosperm and bran by PEN or CEL were needed to make good dough and baking properties.

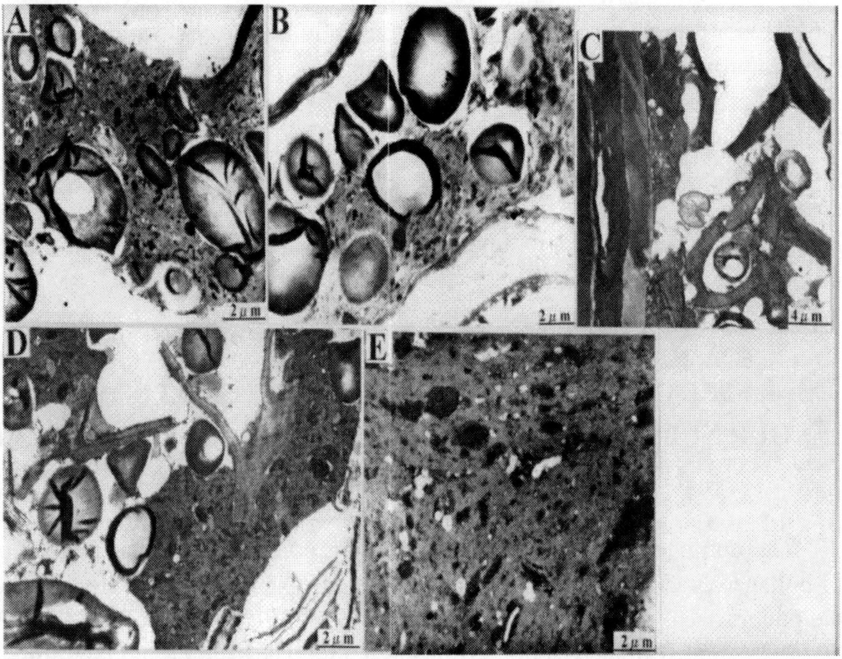

A, Hermes; B, N61; C, NA-1; D, NA-5; E, NA-8. Abbreviations are the same as in Table 1-1-1.

Figure 1-1-2. TEM of wheat dough mixed in a home baker.

Table 1-1-2. Characteristics of various graded flours

Sample	General analysis				Amounts of various components							
	Moisture (%)	Ash (%)	Protein (%)	Dietary fiber (%)	S (%)	Cl (%)	K (%)	Ca (ppm)	Fe (ppm)	Ni (ppm)	WSP (%)	DS (%)
Hermes	11.8	0.4	11.8	1.9	0.17	0.08	0.15	300	29.9	1.6	0.73	15.29
CW	12.3	0.2	12.3	2.1	0.17	0.05	0.10	250	19.3	1.3	0.74	15.47
C-1	11.1	4.3	16.1	32.8	0.27	0.10	1.04	1030	270.2	2.1	0.59	16.54
C-5	9.1	1.0	13.9	5.7	0.20	0.08	0.30	410	48.3	1.7	1.09	39.06
C-8	6.9	0.5	8.1	3.8	0.14	0.07	0.15	270	31.8	1.6	0.80	27.95

C-1, C-5 and C-8 were 100-90 %, 60-50 % and 30-0 % layers of the 1CW whole grains, respectively. WSP, water-soluble pentosan; DS, damaged starch.

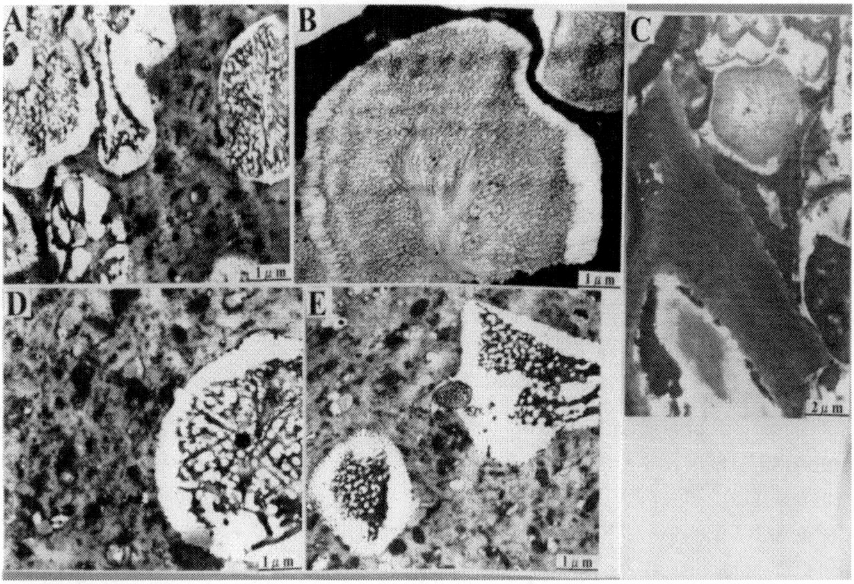

A, Hermes; B, NB-1; C, NA-1; D, NA-5; E, NA-8. Abbreviations are the same as in Table 1-1-1.

Figure 1-1-3. TEM of dough after baking in a home baker.

4. Characterization of Hard-Type Polished-Graded Flours

Graded flour was prepared from the hard-type wheat grain 1CW by the polished-grading method and the effect of the flour quality on the dough and baking properties was studied. The graded flours contained large amounts of ash or dietary fiber as compared with a conventionally milled hard-type 1CW flour (CW) (Table 1-1-2). These graded flours also had larger amounts of K, Ca and Fe than CW. All the graded flours, except for the outermost layer, distinctly increased in water-soluble pentosan as compared with CW (Table 1-1-2). Moreover, they contained higher β-amylase or diastatic activities and also higher amounts of damaged starch than CW. The specific volumes of the bread made from the graded flours were distinctly smaller than that of CW. None of the graded flours alone had a favorable effect on the bread firmness. For the viscoelastic properties, all parameters of the graded flours tested distinctly higher than those of CW. Microscopic observations of the graded flour dough alone indicated that the co-existence of bran could not produce a favorable gluten matrix, resulting in poor quality bread.

These results suggested that graded flour alone could not make dough with the desired baking properties, because of the presence of high amounts of diastatic activity, damaged starch, dietary fiber or wheat bran.

5. Effect of Polished-Graded Hard-Type Wheat Flour Substitution for Commonly Milled Wheat Flour on the Breadmaking

The effect of substituting polished-graded hard-type wheat flour for commonly milled hard-type wheat flour on the properties of dough and bread was studied. The CW was also prepared and breadmaking was carried out by the AACC method. Thirty (30 CW) or 50% (50 CW) graded-flour substitution for CW was tested for baking using the 70~40 % layer fractions of the whole grain.

The addition of pentosanase (PEN) or sucrose fatty ester (SE) to 30 CW distinctly increased the loaf volume or softness of the bread crumbs due to the small size of the gas cells and retarded retrogradation. Moreover, 30 CW with PEN or SE increased the enthalpy value of the starch. On Farinogram, the addition of SE to 30 CW decreased water absorption, arrival and development times, while it increased the stability time as compared with that of CW alone. The viscoelasticity of 30 CW with SE showed larger values for all parameters than those of CW. The ratio between resistance and extensibility of the 30 CW and SE dough increased significantly (after 90 and 135 min) as compared with CW on the extensogram. The substitution with graded flour alone or its combination with PEN or SE increased the rate of gas generation. Microscopic observation of the 30 CW dough with PEN or SE distinctly accelerated formation of the gluten network. The extensible gluten material could entirely cover the starch granules.

From these results, the addition of graded flours and SE to CW might increase maturity of the dough. Furthermore, these additions were considered to control dispersion of the water in the dough and improve the dough and baking properties when compared with those of CW alone.

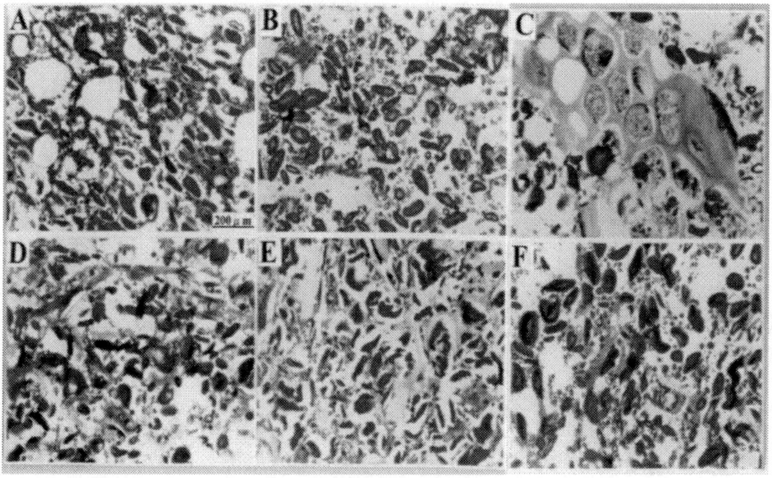

A, Hermes; B, N61; C, NB-1; D, NA-1; E, NA-5; F, NA-8.
Abbreviations are the same as in Table 1-1-1.

Figure 1-1-4. Light microscopic observation of dough stained with PAS and light green.

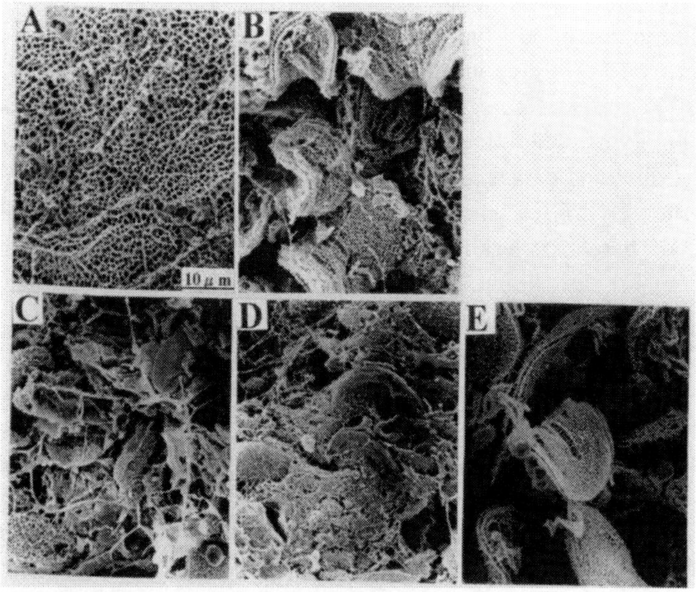

A, Hermes; B, N61; C, NA-1; D, NA-5; E, NA-8.
Abbreviations are the same as in Table 1-1-1.

Figure 1-1-5. Scanning electron microscopic observation of bread baked in an automatic baker.

Table 1-1-3. Effects of pentosan obtained from graded flours on baking properties

Sample	Amount of pentosan added (%)	Specific volume (cm^3/g)	Storage days 0	1	2	3
Hermes	0	4.25	16.7	28.7	41.3	43.3
N61	0	2.73	70.3	134.3	193.7	220.0
WSP 0.2	0.2	3.79	21.3	52.7	78.0	104.3
WSP 0.5	0.5	3.84	19.3	49.0	69.0	89.0
WSP 1.0	1.0	4.38	18.0	40.0	60.0	80.0
WISP 0.5	0.5	2.77	83.3	148.0	202.3	225.7
WISP 1.0	1.0	3.84	20.7	39.7	60.3	83.3
WISP 5.0	5.0	4.07	31.7	90.7	128.7	174.0
WISP 10.0	10.0	3.96	15.7	45.0	67.3	85.7

* Firmness ($10^2 Nm^{-2}$); WSP, water-soluble pentosan; WISP, water-insoluble pentosan.

6. Flour Quality and Pentosan Prepared by Polishing Wheat Grain on Breadmaking

Whole grains of soft-type wheat were polished from the surface layer to the center by 10% of the total weight using a modified rice-polisher. Qualities of classified-graded flours and effect of pentosan in the graded flours on breadmaking were determined. All the graded flours contained larger amounts of dietary fiber and damaged starch, and showed higher diastatic activity than a conventionally milled wheat flours of Hermes or N61. All kinds of graded flours made harder or less moist dough from microscopic observations than Hermes or N61 did (Figure 1-1-4). The gluten matrix of graded flour dough was distinctly not continuous in the presence of large dietary fiber, and the gluten sheet was clearly more discontinuous than those of Hermes and N61. Also, the gelatinization of starch in graded flours during baking appeared to be more incomplete than that observed with Hermes or N61 (Figure 1-1-5). However, additions of water-soluble and -insoluble pentosans obtained from the innermost fraction of graded flours to N61 significantly improved the bread qualities (Table 1-1-3). The improvements of bread quality and maturity of dough might be caused by the large amounts of pentosans and damaged starch in graded flours.

7. Characterization of Pentosan of Various Graded Flours

All the soft-type graded flours contained larger amounts of water-soluble pentosan (WSP) and water-insoluble pentosan (WISP) than those of Hermes and N61. The substitution of the graded flours for N61 increased the amounts of WSP, while the ratio of Ara/Xyl (arabinose to xylose) of WSP decreased. The additions of PEN and CEL could decrease the Ara/Xyl of WSP. Hydrolysis with PEN or CEL increased the solubility of WISP in 50 NA (50% soft-type graded flours substitution for N61) at 50 % or 10 %, respectively, as compared with 50 NA alone. And all the hard-type graded flours had the larger amounts of WSP and WISP similar to the soft-type graded flours. The additions of PEN and SE to 30 CW (30% hard-type graded flours substitution for CW) could increase the solubility of WISP by about 18 % and 5 %, respectively, compared to 30 CW alone. And, the addition of PEN showed the lower ratio of Ara/Xyl of WSP than those of 30 CW and 50 CW alone, whereas the addition of SE slightly increased Ara/Xyl of and viscosity of WSP. The PEN hydrolysis of water-insoluble arabinoxylan (WISAX) decreased molecular weight of its polysaccharide, contributing to make a good dough property. In contrast, the addition of SE to 30 CW affected the WISP entangled in the doughs with the high hydrophilic property, resulting in the looseness of structure of WISAX.

Ara/Xyl of WSP and WISP isolated from NA-8 (the soft-type graded flour from 30-0 layers of whole grain) was lower than that of Hermes. And, in NA-8 the WSP has lowly branched WSAX, while the WISP, highly branched WISAX than those in N61. The WSP and WISP of NA-8 included the larger amount of ferulic acid, and also the WSP made protein foam more stable against heating than those of Hermes and N61.

When WSP and WISP from NA-8 were hydrolyzed by PEN, the degrees of chain length in final products were distinctly decreased. The PEN hydrolysis of WISPs from N61 and NA-8 made them soluble in water by around 65 %. The WSP formed from enzyme hydrolysis of WISP had favorable effects on the breadmaking. Therefore, the substitution of graded flours and PEN for N61 improved breadmaking by the presence of WSP rather than WISP. For the case of corn arabinoxylan, the addition of AX of a high-molecular weight and low ratio of Ara/Xyl improved the dough and baking properties, as compared with that of AX of a low-molecular weight and high ratio of Ara/Xyl. These results indicated that the improvement of graded flours for breadmaking was caused by large amounts of ferulic acid and low Ara/Xyl, which are related to the gel-forming ability of WSP and to the presence of heat stable foam. This stable foam accelerated the extensibility and maturity of the dough, resulting in the production of end products of a good quality. In addition, the combination of graded flours and PEN, the presence of enough amount of WSP, and degradation of WSP and WISP having low Ara/Xyl make the dough extensible and viscous, resulting in favorable end products.

8. General Discussion and Conclusion

Many non-starch polysaccharides, such as β-glucan, hemicellulose, pentosan and gluco-, as well as galacto-mannan are well known to exist in wheat grain. In most cases, pentosans in cereals are linked to protein as a kind of glycoproteins, which are composed of large amount of pentose, small amount of hexsose, and only a little of protein. The nomenclature of these substances in the field of cereal chemistry is somewhat ambiguous, so no district definition has been made between pentosans, hemicelluloses and glycoproteins. The nomenclature in this study was based on the suggestion by Meuser and Suckow [1988]. Wheat flour has about 2-3 % of pentosan, and about 25 % of the pentosan is soluble in water (WSP). WISP means alkali-soluble pentosan or hemicellulose and its structure is essentially similar to that of WSP, except for the differences in the degree of branching and molecular weight. WSP prepared from wheat flour is usually reported to contain some hexoses, protein and phenol as impurities. It could be classified into two fractions, judged from solubility in saturated ammonium sulfate solution or in 80 % ethanol. The precipitated fraction was AX complexed with protein. The soluble fraction was arabinogalactan (AG) covalently linked to a protein. AX has been proposed to take up an extended left-handed three-fold helix which has a long main chain of β-(1-4)-linked D-xylopyranose units having side chains of L-arabinofuranose linked at the position 2 or 3 of xylose residue. The side chains are responsible for the water solubility of AX; the backbone xylan of which precipitates easily when the arabinose residue is hydrolytically liberated from the molecule [Neuko et al., 1967]. These AX and AG fractions are water-soluble glycoproteins named arabinoxylan-protein complexes (AXPC) and arbinogalactan-protein complexes (AGPC), respectively. These substances are also referred to as WSP [Pomeranz, 1988a].

AXPC contains a small amount of a phenolic component (2.0 and 0.01 mg/g for AXPC and flour, respectively) that is labile to oxidation and has been identified as ferulic acid. Ferulic acid or diferulic acid are covalently linked to WSP and WISP. The diferulic acid has been reported to serve as coupling unit to an insoluble gel network during oxidative gelation of WSP [Meuser and Suckow, 1988]. Especially, a wheat flour of a good quality contains higher amounts of AXPC and polyphenols than that of a poor quality. The higher content and degree of oxidative gelation of AXPC in a good quality flour resulted in a higher viscosity than that from a poor quality flour.

The addition of AXPC of a good quality flour to gluten-starch mixture shows more favorable effects on the improvement of dough and baking properties, than that of a poor quality flour. Furthermore, the addition of oxidants increased the effects. The oxidative cross-linkage of WSP containing glycoproteins has an evident effect on dough properties because of a high water-holding capacity and gel formation. WSP might play a regulatory role with respect to water dispersion in the dough during breadmaking. Because WSP neither coagulates through heat treatment (as is the case with protein), nor retrogrades through cooling and storage (the case with starch), it has positive effects on keeping bread fresh. WSP can hold water even during baking processes, and then water is continuously supplied to the starch granules, thereby, the rate of retrogradation is retarded. In addition, WSP has an ability to retain gas in the dough with a formation of 'membrane' [Meuser and Suckow, 1988]. From these facts, WSP has been considered to have positive effects on the baking properties, such as loaf volume, crumb structure and staling of bread.

For the case of WISP, it might be swollen and the water-holding capacity is 3.5-6.3 and 6.7-9.9 times of its weight for WSP and WISP, respectively. The high water absorption in the dough would depend on the conformation of these pentosan molecules. This change would be derived from the association between carbohydrates and proteins [Meuser and Suckow, 1988]. WISP of wheat is primarily composed of AX, but contained a small amount of hexoses, such as mannose and glucose, which can form mannan or glucomannan as polysaccharides [Meuser and Suckow, 1988]. The insolubility of WISP is explained from the strong physical entanglement of WISP caused by the higher degree of branching than WSP, together with the presence of phenolic acid in the tailings [Pomeranz, 1988a]. But, reasons for the insolubility have not been fully understood. Total content of phenolic acid which was calculated as ferulic acid was found to be 1.2 and 0.4 mg/g of tailings and flour, respectively.

WISP has also been reported to improve the loaf volume, freshness, or staling, crumb structure of the bread and elasticity of the dough [Meuser and Suckow, 1988; Pomeranz, 1988a]. In addition, WISP was found to be a binding agent suitable for making bread from non-gluten flour. Regarding the retrogradation of starch in the dough, the effects of pentosans have been reported as follows: WSP appeared to affect the solubility of amylopectin in starch, and WISP retarded the retrogradation of both amylose and amylopectin. The rate of starch setback was slowdown more obviously by WISP than by WSP [Pomeranz, 1988a].

These results lead to a conclusion that WSP and WISP in wheat flour influence the rheological behavior of doughs and texture of bakery products. This effect was possibly due to the high water-holding capacity of WISP as well as their swelling ability, to the ability of WSP to form highly viscous aqueous solutions, and to the capacity of glycoproteins to increase the viscosity under the influence of oxidants. In this study, authors examined dough and baking properties of polished-graded flour. The graded flour alone could not show good dough and baking properties regardless of the soft- or hard-type wheat grain, and these graded flours had almost the same flour quality. All the graded flours contained larger amounts of dietary fiber, ash and minerals than conventionally milled flours. In addition, the formers showed a higher diastatic activity and contained a larger amount of damaged starch. Since the dietary fiber (WSP and WISP), diastatic activity and damaged starch might absorb an excess amount of water and promote the degradation of starch in the dough, they might give rise to the poor baking qualities. However, these factors are nutritionally effective, and they might be also expected to bring about the enough fermentation on the breadmaking. Even though these unfavorable materials for breadmaking contained in the flour, the graded flours have been used because of the presence of large amount of dietary fiber. The poor dough and bread properties might be caused by the high water-holding capacity of WSP and WISP, resulting in the deficiency of water to make a good gluten structure. Moreover, large amounts of damaged starch and high diastatic activity in the graded flours made the gluten structure unstable, which leads to retard the extensibility of gluten and gelatinization of starch during subsequent baking processes. However, when the graded flours were partially substituted for the conventionally milled flours, the dough properties could be improved by the increase in amounts of damaged starch and diastatic activity in substituted flours. As a result, the maturity and gas generation in doughs were increased during fermentation.

Additives such as HEM, PEN, CEL and SE improved dough and baking properties. PEN could hydrolyze WISP and then produced the WSP. Thus formed WSP might change the gluten structure to be quite extensible, providing the expansion of the dough during fermentation and baking processes. WISP has been considered to retard the formation of good gluten structure by the high water-holding capacity. However, when PEN was added to the substituted flours, the part of WISP might be decomposed and solubilized, resulting in construction of good gluten structure. In contrast, SE with a high ability of emulsification increased the viscosity of water-extracts containing WSP. Moreover, SE might control the distribution of water to protein and starch by the high surface activity, even if WISP prepared from the graded flours were added. Consequently, the substituted flours with SE were inferred to increase the suitable hardness of dough, caused by the entanglement of WISP, and also to prevent the dough from escaping the generated gas during fermentation.

Although effects of PEN and SE on breadmaking were apparently different, the quality of end products was quite similar. This fact suggests that the mechanisms of baking process are quite complicated, because many kinds of constituents in the flour and ingredients added change individually during the baking processes, and also interact each other. Nevertheless, the properties of WSP or WISP on the breadmaking as reported previously were sure to associate strongly with the qualities of graded flours.

Therefore, the combination of graded flours with enzyme and emulsifier improved not only the dough and baking properties but also the nutrition in the end product. In addition, these improvements might be due to characteristics of WSP and WISP as described above. However, further extensive studies are needed for polished-graded wheat flours to make clear not only flour characteristics but also the dough and baking properties, together with the application of this flour for other food ingredients. Consequently, the practical use of graded flours for our daily diet as a new foodstuff will be desired.

REFERENCES

Cornell, H. J., and Hoveling, A. W. (1998). The wheat kernel. In *Wheat: Chemistry and Utilization* (pp. 3-4). Switzerland: Technomic Publishing Co. Inc.

Meuser, F., and Suckow, P. (1988). Non-starch polysaccharides. In *Chemistry and Physics of Baking,* Blanshard, J. M. V., Frazier, P. J., and Galliard, T. (Eds.) (p. 47, 50, 51, 54, 59) Australia: Royal Soc. Sci.

Nagao, S. (1996). In *Science of Wheat* (pp. 37-39). Tokyo: Asakura Publishing Co. (in Japanese).

Neuko, H., Providoli, L., Gremli, H., and Hui, P. A. (1967). Recent investigations on wheat flour pentosans. *Cereal Chemistry,* 44, 238-244.

Pomeranz, Y. (1988a). Microscopic structure of the grain. In *Wheat*: *Chemistry and Technology*, I, Pomeranz, Y., 3rd Ed. (pp. 47, 54, 110-119, 118-138, 139-147, 341, 352-353). St. Paul, M. N., US: A. A. C. C., Inc.

Pomeranz, Y. (1988b). Wheat flour milling. In *Wheat: Chemistry and Technology,* II, Pomeranz, Y., 3rd Ed. (pp. 8-17). St. Paul, M. N., US: A. A. C. C., Inc.
NinoA

In: Advances in Cereal and Pseudocereal ... ISBN: 978-1-62618-347-6
Editors: N. Morita, P.V. Hung et al.

Chapter 2

SOURDOUGH BREAD

Tomoko Maeda[1], Ji Hyun Kim[2] and Naofumi Morita[3]
[1]Department of Science, Technology, and Human Life,
Graduate School of Education, Hyogo University of Teacher Education,
Shimokume, Kato, Japan
[2]Lab. of Food Chemistry, Graduate School of Life and Environmental
Sciences, Osaka Prefecture University, , Sakai, Naka-ku, Osaka, Japan
[3]Department of Food Packaging Technology, Toyo College of Food
Technology, Mimami-Hanayashiki, Kawanishi, Hyogo, Japan

ABSTRACT

To make the best use of polished flours for processed foods, sourdoughs were prepared using three fractions of polished flours C-1 (100-90%), C-5 (60-50%) and C-8 (30-0%) and their sourdough properties and bread qualities were studied. The pH and total titratable acidity levels and the buffering capacityof sourdoughs made from polished flours were significantly different from those of the control sourdough made with No. 1 Canada Western Red Spring (CW), and these flours were sufficient for providing better quality sourdough breads, compared with CW. The growth of lactic acid bacteria and yeast in polished flour sourdoughs was significantly accelerated during fermentation, compared with that in CW sourdough. The higher maturation of polished flour sourdough altered the hardness of the mixed dough. Substitutions of sourdoughs made from polished-flour fractions C-5 and C-8 in place of wheat flours significantly increased the loaf volume, and the breadcrumbs containing sourdoughs made from the C-5 and C-8 fractions were significantly softer than that of CW sourdough. The qualities of polished flours, such as larger amounts of damaged starch and dietary fiber or higher maltose values, and the suitable acidity or good buffering capacity caused by the bran fraction were effective for better growth and activity of the yeast in the dough during fermentation. The volatile flavor compounds identified in C-5 and C-8 were similar to those found in CW. However, some compounds that have a connection with both oxidation products from unsaturated fatty acids and metabolites from a number of species associated with moldy grains were more abundant in C-5 and C-8 than in CW. The sourdough-method increased the proportions of acids and aldehydes in the crust but decreased the levels of alcohols and methoxybenzenes identified in the outermost fraction compared to the straight-common-method. In addition, the sourdough-method lowered the amounts of some compounds related to the oxidation products from unsaturated fatty acids andmetabolite species obtained from moldy grains, compared with the straight-common-method. Therefore, the sourdough-method, including lactic acid fermentation, would be a suitable baking procedure that uses polished flours based on flavor properties. Sourdough bread using polished flours obtained from the middle and innermost fractionsmight produce the taste and texture qualities that best satisfy consumers' bread preferences.

1. Application of Polished-graded Wheat Grains for Sourdough Bread

Wheat is one of the most predominant crops consumed in the world, and it has been widely used in various processed foods. High quality wheat flours, with properties such as a white color, glossy appearance and fine size, are essential for producing processed foods. As a result, milling technology and its precise processes have been optimized to completely remove bran or germ from the whole-wheat grains. Therefore, many nutrient components such as vitamins, minerals and dietary fibers in the bran and germ are usually discarded during the conventional milling process. However, health-conscious consumers want to eat or reuse these nutritious by-products for their daily diet. Instead of the conventional milling method, a new polishing-grading method has been developed, and the milling value of polished flours and its applications to bread and cookie making have already been studied and reported (Maeda et al., 1999, 2001, 2004; Maeda & Morita, 2000, 2001, 2003). The partial substitution of polished flours for common flours increased the loaf volume and retarded the staling of bread while only requiring some modifications in the baking procedures and combined additions of enzymes or emulsifiers. However, polished flours are not yet used routinely in processed foods, similar to conventional whole-grain flours, although consumers might begin demanding more natural, safe and healthy foods than those produced with conventionally milled flour.

Sourdough bread made from rye flour is one of the brown breads that contain bran or germ, and it has been increasingly eaten by health-conscious consumers who are interested in leading healthy lives. The *Lactobacillus* in sourdough generally requires various types of nutrient components for its fermentation. Therefore, multiple bran-hydration methods (Salmenkallio-Marttila et al., 2001; Degutyteet et al., 2002) have been employed. When improved dough and bread properties are utilized, the bread baked from sourdough can yield a better flavor than those prepared using commercial baker's yeast (Brümmer & Lorenz, 1991; Röcken, 1996). In particular, the type of flour is a major factor in evaluating acidity, including the properties of hydrogen potential (pH), total titratable acidity (TTA) and lactic and acetic acid producing power. The relationship between the flour and acidity has been widely investigated using normal whole-wheat and rye flours (Hansen & Hansen, 1994). Because the present polished flours also contain large amounts of pentosans, damaged starches and dietary fibers aside from the nutrient components, these components could be applied to the sourdough, replenishing the rye flour with the valuable components that were removed during the milling process. Hence, these applications are expected to produce favorable dough fermentation conditions and a delicious bread flavor, as rye flour does (Kim et al., 2005; Maeda et al., 2008, 2009). In addition, the present polished flours were prepared by a different polishing method than the method used for conventional whole-grain flours. They are expected to provide sufficient sourdough properties similar to those of normal whole-grain flours while also increasing the loaf volume and retarding the staling of breadcrumbs. Therefore, polished flours were used for preparing sourdough to increase the amount of polished flours in the daily diet, and the effects of using these flours in sourdough bread are described.

Flours obtained by a specific polishing process were used to prepare sourdough and bread. Three fractions of C-1 (100-90%), C-5 (60-50%) and C-8 (30-0%) were studied (Kim et al., 2005). The pH, TTA levels and buffering capacity of sourdoughs made from polished flours were significantly different from those of the control sourdough with CW, and these parameters were sufficient for providing better qualities in sourdough breads compared to CW (Fig. 2-1).

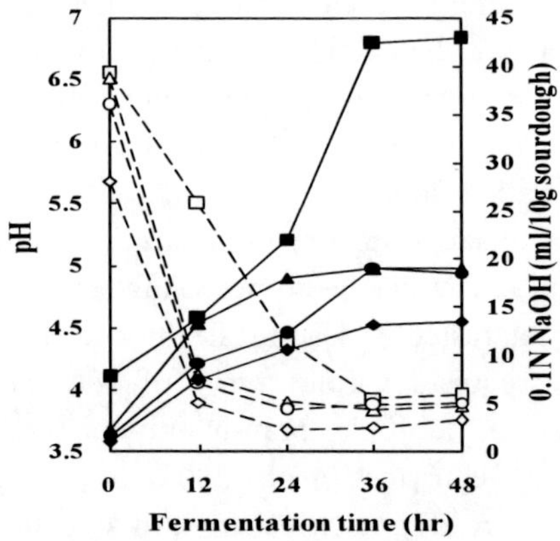

Fig. 2-1. Time course changes in the pH and total titratable acidity (TTA) of various sourdoughs during fermentation (adapted from Kim et al., 2005).
pH: ◊, CW; □, C-1; Δ, C-5; ○, C-8. TTA: ♦ CW; ■, C-1; ▲, C-5; ●, C-8.
CW, common hard-type wheat flour; C-1, polished flours from 100-90%; C-5, 60-50%; C-8, 30-0% of whole-wheat grains.

The growth of lactic acid bacteria and yeast in sourdoughs made from polished flours were significantly accelerated during fermentation compared to CW sourdough. The higher maturation of polished flour sourdoughs softened the hardness of the mixed dough (Fig. 2-2).

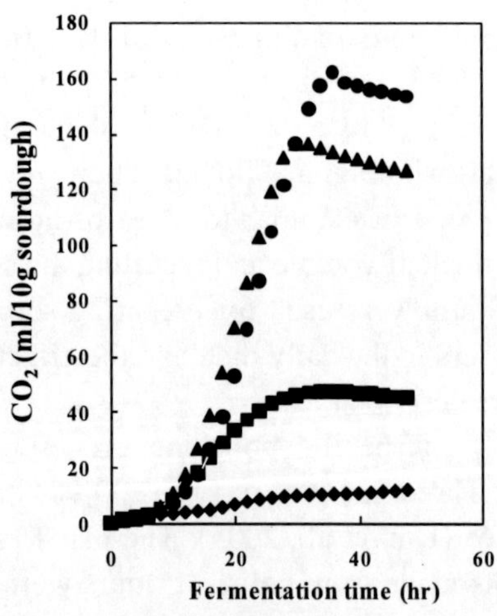

Fig. 2-2. Effects of polished flours on the gas generation of sourdoughs during fermentation (adapted from Kim et al., 2005).
♦ CW; ■, C-1; ▲, C-5; ●, C-8. Abbreviations are the same as in Fig. 2-1.

The intricate network of honeycomb-structure gluten and the uneven surface of the starch granules were distinctly observed in SEM images. Substitutions of C-5 or C-8 sourdoughs for CW significantly increased the loaf volume and the softened breadcrumbs more than the CW sourdough (Fig. 2-3).

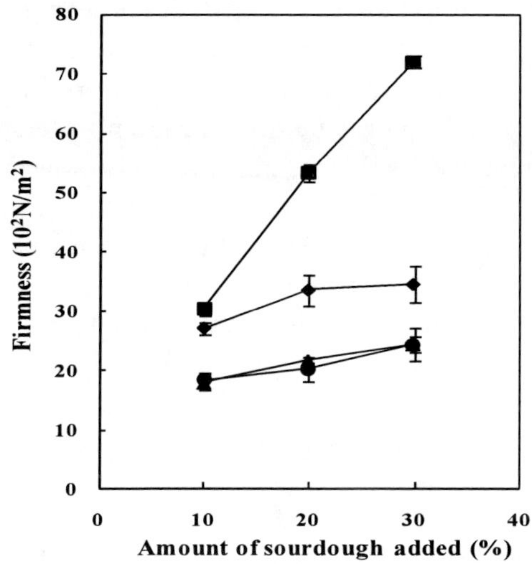

Fig. 2-3. Effects of various sourdoughs on the firmness of breadcrumbs after baking (adapted from Kim et al., 2005).
♦ CW; ■, C-1; ▲, C-5; ●, C-8. Abbreviations are the same as in Fig. 2-1.

The flour qualities of the polished flours, such as suitable acidity and good buffering capacity caused by the bran fraction, were effective for better growth and longer yeast life in the dough during fermentation. Therefore, the application of polished flours for sourdough bread would improve the rheological properties of dough and bread compared with CW sourdough.

When the polished-flour sourdoughs were substituted for wheat flours, the rheological properties during fermentation and the flour qualities such as larger amounts of damaged starch and dietary fiber, higher maltose value and enzyme activity were effective for producing the proper acidity required for the growth of microbial cells in the dough, resulting in accelerated fermentation and favorable baking results.

2. Analysis of Volatile Compounds in Polished-Graded Wheat Flours Using Headspace Sorptive Extraction

Recently considered to be a new foodstuff, polished-graded wheat flours are prepared from whole-wheat grains using a gradual polishing-grading method. Because the polishing procedure was processed without debranning and removing the germ, a procedure differing from the conventional milling method reported previously (Maeda et al., 1999, 2001; Maeda & Morita, 2000, 2001, 2003), the prepared flours have various benefits, including high nutritional and functional values, compared to commonly milled wheat flours. We have reported that the polished-graded wheat flours improved the rheological and physicochemical properties of dough and bread when some additives, including enzymes and polysaccharides, were added to the ingredients (Maeda et al., 2004; Kim et al., 2005, 2006). However, the presence of off-odors derived from the bran and germ in polished-graded wheat flours may hinder their practical application in processed foods such as breads, noodles and cookies. To resolve this problem, it is necessary to identify the volatile flavor compounds in polished-graded wheat flours. In addition, it is important to make clear the causes of volatile flavor compounds (off-odors) in polished-graded wheat flours. Although cereal grain, especially wheat, is an important ingredient in the human diet, there have only been a few reports concerning the volatile compounds in raw cereal grain or flour (McWilliams & Mackey, 1969; Hougen et al., 1971) until recently. Progress in sample preparation and detection methods has accelerated the research on flavor compounds found in wheat and other grains. In 1969, Mcwilliams and Mackey (1969) isolated flavor compounds from wheat grains by steam distillation under vacuum and by headspace vapor under atmospheric pressure. On the other hand, Wasowicz and Kamiński (1988) used simultaneous distillation/extraction (SDE) for the isolation of volatile flavor components in sound and musty wheat grains. However, these methods involving extraction with organic solvents may lead to serious errors in the volatile flavor analysis of cereal grain, due to the long extraction time and loss of flavor components during sample concentration, either under nitrogen gas or in a rotary evaporator under reduced pressure. The merits of headspace sampling for the volatile compounds associated with flavor could be the high sensitivity, high extraction rate of volatile compounds and high purity of the extracts, compared with those obtained by solvent extraction or distillation methods (Chang et al.,

1995). Thus, dynamic headspace (purge and trap) methods have been mainly used to identify volatile compounds in whole grains with the various volatile flavors (Seitz & Sauer, 1994). Seitz et al. (1999) attempted to isolate volatile compounds in whole and ground grains by combining supercritical fluid extraction (SFE) with the dynamic headspace method (DHM), and they reported that, compared to DHM alone, the combination method of SFE and DHM was more suitable for isolating aldehydes, 2,3-butanediols, acetic acid and chloro- and bromo-methoxybenzenes, which are associated with the musty odors in grain.

Recently, a method using a magnetic stirring bar coated with poly-dimethylsiloxane (PDMS) was developed as a new technique for extracting volatile flavors (Baltussen et al., 1999; Salinas et al., 2004). This magnetic stirring bar is sold under the name "Twister", and the technique is known as stir bar sorptive extraction (SBSE). The extraction mechanism and advantages of SBSE are similar to solid-phase micro-extraction (SPME), except for its higher ratio of coating phases, which therefore result in up to 100 times lower detection limits than SPME (Salinas et al., 2004). Compared with the other extraction methods, SBSE has many advantages, such as solvent-free, rapid sampling with low cost, easy operation and sensitivity. Thus, this technique has been used with success in the analysis of volatile flavor components found in various foodstuffs, such as grapes, wines, coffees, teas and juices (Salinas et al., 2004). However, to the best of our knowledge, this technique has not been applied to solid foodstuffs, such as cereal grain and its by-products.

Like the coated SPME fiber, a Twister could adsorb the headspace over a liquid or solid sample and directly within an aqueous sample. Because volatile compounds can be concentrated without interference by matrix components compared with SBSE, headspace SBSE (i.e., headspace sorption extraction, HSSE) may be ideally suited to analyze the volatile flavor compounds of solid masses in foodstuffs with low volatility, such as wheat grain, flour, and bread. The main aim of this study is to identify and characterize the volatile flavor compounds obtained by the HSSE method in polished-graded wheat flours, as well as to investigate the causes of their off-odors.

The HSSE method was used to determine the volatile flavor compounds of the polished-graded wheat flours and commonly milled wheat flour (CW). The polished-graded wheat flours of three fractions, C-1 (100-90%), C-5 (60-50%) and C-8 (30-0%), as well as CW obtained from a hard-type wheat cultivar 1CW were used in the present study (Maeda et al., 2008). A total of 48 volatile flavor compounds present in all of the samples were commonly found in wheat grain or flour. The major volatile compounds in C-1 included methoxybenzene (i.e., 1,4-dimethoxybenzene, 1,2,4-trimethoxybenzene, and 1,2-dimethoxybenzene), associated with musty odors, and nitrogen compounds (i.e., trimethylamine, 3(4)-methyl pyridine, and 3-methoxy pyridine). The volatile flavor compounds identified in C-5 and C-8 were similar to those in CW (Table 2-1).

Table 2-1. Composition of the volatile flavor compounds found in polished-graded wheat flours and commonly milled wheat flour (adapted from Maeda et al., 2008)

Sample	Methoxybenzene	Aldehydes	Alcohols	Hydrocarbons	Bases	Others	Ketones	Esters
				(%)				
CW	1.92	16.54	42.62	25.04	0.72	4.41	6.34	2.41
C-1	67.51	0	1.82	21.07	4.51	4.85	0.24	0
C-5	1.32	23.01	42.49	17.07	0.56	4.59	7.97	2.67
C-8	1.02	29.09	43.32	14.11	0.29	2.76	6.66	1.87

CW, common hard-type wheat flour of 1CW; C-1, C-5 and C-8 were polished-graded wheat flours obtained from the 100-90%, 60-50% and 30-0% layers of 1CW whole-wheat grains, respectively.

However, some of the compounds that have a connection with both oxidation products from the unsaturated fatty acids and metabolites of a number of species associated with moldy grains were more abundant in C-5 and C-8 than in CW. Furthermore, hexanal, methanol, ethanol, hexanol, 1-octen-3-ol, 3-octen-2-one, (E,Z)-3,5-octadien-2-one and (E,E)-3,5-octadien-2-one were considerably abundant in C-5 and C-8.

The volatilities of polished-graded wheat flours were noticeably different from those of CW. C-1, the outermost layer of whole-wheat grains, contained abundant amounts of methoxybenzene and nitrogen compounds. However, C-5 and C-8 contained large amounts of some of the volatile compounds derived from the oxidative degradation of unsaturated fatty acids and their metabolites. As for the free fatty acids content, polished-graded wheat flours, except for C-8, showed 2.0-3.0-fold greater amounts of free fatty acids than CW (Table 2-2).

As a result, no distinct relationships were observed between the volatile flavor compounds, known as off-odor, and free fatty acid contents.

Table 2-2. Contents of free fatty acids in polished-graded wheat flours and commonly milled wheat flour (adapted from Maeda et al., 2008)

Free fatty acid		Various wheat flours[a]			
		CW	C-1	C-5	C-8
$C_{12:0}$	Lauric acid	-	1.00a	0.97a	1.10a
$C_{14:0}$	Myristic acid	1.00a	1.07a	1.13a	1.00a
$C_{16:0}$	Palmitic acid	7.93b	18.97c	20.67cd	6.77a
$C_{18:0}$	Stearic acid	1.33a	1.50ab	1.73b	1.17a
$C_{18:1}$	Oleic acid	6.53a	18.53b	16.47b	5.40a
$C_{18:2}$	Linoleic acid	25.90b	66.90d	29.40c	21.07a
$C_{18:3}$	Linolenic acid	2.37a	10.30b	2.17a	2.27a
$C_{22:0}$	Behenic acid	1.27a	1.37a	1.03a	-
Others		1.40ab	1.67b	1.73b	0.97a

Values followed by different letters in the same row are significantly different according to Duncan's multiple range test ($P<0.05$). Unit, 10^{-2}mg/g flour.
[a] CW, common hard-type wheat flour of 1CW; C-1, C-5 and C-8 were polished-graded wheat flours obtained from the 100-90%, 60-50% and 30-0% layers of 1CW whole-wheat grains, respectively.

3. Analysis of Volatile Compounds in Polished-Graded Wheat Flour Bread Using Headspace Sorptive Extraction

Sourdough fermentation is a classical breadmaking process that utilizes the first known bread leavening method available prior to the discovery of yeast in beer brewing (Röcken &Voysey, 1995). Sourdough is defined as a dough that includes microorganisms such as lactic acid bacteria and yeasts, which are active or can be reactivated in sourdough or sourdough starters (Clarke et al., 2004). A great part of these advantages is accelerated by the acidification of sourdough; this includes gas retention in the gluten network, water binding to gluten and starch granules, swelling of pentosan, solubilization of the phytate complex by endogenous phytases, and the prevention of spoilage by lowering the pH during fermentation (Gobbetti, 1998). Acidification of sourdough during the breadmaking process improves the rheological properties of the dough, leading to more cohesive crumbs with uniform structures. Additionally, the shelf life of the bread is extended by delaying the growth of mold (Brümmer et al., 1984; Collar et al., 1994). In addition, the acidification of sourdough promotes the proteolysis of gluten subunits and the hydrolysis of starch during dough fermentation, thus changing the physicochemical properties of the dough and delaying the hardening and staling of the bread (Corsetti et al., 1998; Salmenkallio-Marttila et al., 2001). Furthermore, the acidification is affected by the flour type, temperature, water content, type or amount of added yeast and lactic acid bacteria, which are all important factors that improve sourdough bread qualities (Brümmer & Unbehend, 1997; Rounaud & Martínez-Anaya, 1997). Recently, polished-graded wheat flours prepared by the polished-graded method have been considered a new foodstuff in terms of functional and nutritional properties. Because the polished flours include bran and the germ of wheat grain, they contain larger amounts of minerals, vitamins and dietary fibers than conventionally milled flours, as reported previously (Maeda et al., 1999; Maeda & Morita, 2003). Moreover, they showed high enzymatic activity and large amounts of water-soluble pentosan (Maeda et al., 1999; Maeda & Morita, 2003). Therefore, the polished flours were found to be a good source of these nutrients, and the use of polished flours as an ingredient for breadmaking is sure to increase the intake of dietary fiber. This benefit may be due to the presence of large amounts of bran and germ in polished flours. In addition, the authors suggested that the sponge-dough method, with a longer fermentation time compared to other baking methods, was suitable for endowing polished flours with better dough effects and bread qualities (Maeda et al., 2004). This result is consistent with those reported by Wehrle and Arendt (1998) who suggested that the lowering of pH levels in the sponge-dough method coupled with a long fermentation time increased the peak temperature of starch melting, indicating the modification of starch properties and changes to the rheological properties of dough. Consequently, the drop in pH levels and modification of starch properties were likely caused by the sourdough properties during fermentation. Additionally, Salmenkallio-Marttila et al. (2001) and Brümmer and Unbehend (1997) reported that the fermentation of wheat bran with yeast and lactic acid bacteria improved the loaf volume, crumb structure and shelf life of bread supplemented with 20%

bran. Therefore, sourdough fermentation was expected to provide the polished flours necessary for breadmaking. The volatile compounds in polished flours have already been analyzed and compared to those of conventionally milled wheat flours (Maeda et al., 2008). As a result, the compounds are quite different, depending on the wheat grain portion, and their composition was also different from that of conventionally milled flours. Therefore, when the polished flours were used for processed foods, each layer could be selected depending on the characteristics of each fraction. Because flavor is a very important contributing factor in the quality of foods, especially breads, we analyzed volatile compounds to confirm the difference between the polished flours and common flours. Therefore, in this study and subsequent experiments, the comparison of volatile compounds in breads between the different baking methods will be described to confirm the flavor quality of the breads, and suitable baking methods for polished flours that improve various processed foods will also be discussed.

Flavor compounds are very important factors in the taste of commercial breads. To determine the characteristics of polished-graded wheat flours for breadmaking, the volatile flavor compounds in the bread samples were determined by the HSSE method, comparing two types of baking methods: the straight-common-method (SCM) and the sourdough-method (SDM) (Maeda et al., 2009). The polished flours of three fractions, C-1 (100-90%), C-5 (60-50%) and C-8 (30-0%), and common flour, CW were used. A total of 87 volatile flavor compounds, including 14 alcohols, 13 aldehydes, 9 ketones, 10 acids, 10 esters, 8 furans, 15 bases, 4 sulfur compounds, 3 methoxybenzenes and 1 hydrocarbon were identified in SCM. Interestingly, ethyl lactate and ethyl linoleate were only found in SDM, with a total of 90 volatile compounds identified in SDM (Table 2-3).

In addition, SDM increased the proportions of acids and aldehydes in the crust, while decreasing the levels of alcohols and methoxybenzenes that were identified as volatile C-1 compounds compared with SCM. Regardless of the baking methods, the amounts of 2-methyl-propanol (iso-butanol) and 2-phenylethanol (β-phenyl-ethyl-alcohol), which are known as favorable flavor compounds in breadmaking, were more abundant in the polished flour breads than in the CW bread. Therefore, all the fractions of polished flours might provide bread qualities that consumers find delicious. SDM also decreased the amounts of certain compounds related to the oxidation products from unsaturated fatty acids and the metabolites of species with moldy grains in the breads. Except for C-1 of the outermost fraction, C-5 and C-8 of polished flours improved the dough and bread qualities by using SDM as reported previously. From these results, through the single utilization of C-1, the outermost fraction was found to be poor for breadmaking, and the SDM with lactic acid fermentation was found to be suitable for the baking procedures that utilize C-5 and C-8 polished flours with bran and germ, compared with SCM.

Table 2-3. Composition of identified volatile compounds contained in the crust and crumbs of breads made by SCM and SDM (adapted from Maeda et al., 2009)

SCM	CW		C-1		C-5		C-8	
	Crust	Crumb	Crust	Crumb	Crust	Crumb	Crust	Crumb
Aldehydes	1.70	1.33	2.53	1.00	1.99	1.85	2.42	1.87
Ketones	2.98	3.58	1.91	1.39	1.09	0.59	2.48	0.66
Alcohols	81.09	85.27	74.84	83.61	75.87	85.58	71.70	75.55
Acids	5.50	3.62	6.57	3.77	16.86	6.89	17.82	17.35
Esters	0.73	0.91	1.80	1.99	0.65	0.97	0.75	0.75
Furans	2.10	0.17	2.87	1.21	0.79	0.62	1.42	0.48
Bases	1.67	0.09	1.08	0.04	0.56	0.05	0.46	0.06
Sulfur compounds	0.46	0.10	0.37	0.17	0.19	0.18	0.15	0.09
Methoxybenzenes	0.02	0.03	1.88	2.19	0.06	0.13	0.01	0.01
Others	3.75	4.9	6.15	5.63	1.94	3.14	2.79	3.18
SDM	**CW**		**C-1**		**C-5**		**C-8**	
	Crust	Crumb	Crust	Crumb	Crust	Crumb	Crust	Crumb
Aldehydes	2.33	1.08	3.26	1.04	2.16	2.04	2.48	1.33
Ketones	2.56	1.88	2.12	0.84	1.97	2.28	2.61	0.97
Alcohols	68.90	80.09	74.46	81.35	74.42	72.34	66.60	67.37
Acids	11.66	10.53	7.62	5.81	14.28	16.23	18.94	23.06
Esters	1.56	1.05	2.42	2.26	1.57	1.71	1.30	0.92
Furans	3.62	0.38	2.50	0.55	2.68	0.56	4.52	0.47
Bases	2.42	tr	1.48	0.05	0.66	0.02	0.95	0.03
Sulfur compounds	0.71	0.14	0.67	0.18	0.27	0.34	0.16	0.09
Methoxybenzenes	tr	tr	0.88	2.22	0.03	0.09	tr	tr
Others	6.24	4.85	4.59	5.7	1.96	4.39	2.44	5.76

CW, common hard-type wheat flour; polished wheat flours C-1 (100-90%), C-5 (60-50%) and C-8 (30-0%) of whole-wheat grains. SCM, straight-common-method; SDM, sourdough-method.

REFERENCES

Baltussen, E., Sandra, P., David, F., & Cramers, C. A. (1999). Stir bar sorptive extraction (SBSE), a novel extraction technique for aqueous samples: Theory and principles. *Journal of Microcolumn Separations*, 11, 737-747.

Brümmer, J. M., & Lorenz, K. (1991). European developments in wheat sourdoughs. *Cereal Foods World*, 36, 310-314.

Brümmer, J. M., & Unbehend, G. (1997). *GetreideMehl und Brot*, 51, 345-349.

Brümmer, J. M., Stephan, H., & Spicher, G. (1984). Backtechnischewirkung von weizenvor-und weizensauerteigen. *GetreideMehl und Brot*, 38, 203-205.

Chang, C-Y., Seitz, L. M., & Chambers, E. (1995). Volatile flavor components of breads made from hard red winter wheat and hard white winter wheat. *Cereal Chemistry*, 72, 237-242.

Clarke, C. I., Schober, T. J., Dockery, P., Sullivan, K., & Arendt, E. K. (2004).Wheat sourdough fermentation: Effects of time and acidification of fundamental rheological properties. *Cereal Chemistry*, 81, 409-417.

Collar, C., Benedito de Barber, C., & Martínez-Anaya, M. A. (1994). Microbial sour doughs influence acidification properties and breadmaking potential of wheat dough. *Journal of Food Science*, 59, 629-633.

Corsetti, A., Gobbetti, M., Balestrieri, F., Paoletti, F., Russi, L., & Rossi, J. (1998). Sourdough lactic acid bacteria: Effects on bread firmness and stalin. *Journal of Food Science*, 63, 347-351.

Degutyte-F.L., Sontag-S. T., & Salovaara, H. (2002).Oat bran fermentation by rye sourdough. *Cereal Chemistry*, 79, 345-348.

Gobbetti, M. (1998). The sourdough microflora: Interactions of lactic acid bacteria and yeasts. *Trends in Food Science and Technology*, 9, 267-274.

Hansen, Å., & Hansen, B. (1994). Influence of wheat flour type on the production of flavour compounds in wheat sourdoughs. *Journal of Cereal Science*, 19, 185-190.

Hougen, F. W., Quilliam, M A., & Curran, W. A. (1971). Headspace vapors from cereal grains. *Journal of Agricultural and Food Chemistry*, 19, 182-183.

Kim, J. H., Maeda, T., & Morita, N. (2005).Application of polished-graded wheat grains for sourdough bread. *Cereal Chemistry*, 82, 144-151.

Kim, J. H., Maeda, T., & Morita, N. (2006). Effect of fungal alpha-amylase on the dough properties and bread quality of wheat flour substituted with polished flours. *Food Research International*, 39, 117-126.

Maeda, T., & Morita, N. (2000). Effects of polished-polished flour substitution to commonly milled wheat flour on the properties of dough and bread. *Journal of Applied Glycoscience*, 47, 1-12.

Maeda, T., & Morita, N. (2001). Effects of quality of hard-type polished-flour on breadmaking. *Journal of Applied Glycoscience*, 48, 63-70.

Maeda, T., & Morita, N. (2003). Flour quality and pentosan prepared by polishing wheat grain on breadmaking. *Food Research International*, 36, 603-610.

Maeda, T., Kim, J. H., & Morita, N. (2004). Evaluation of various baking methods for polished wheat flours. *Cereal Chemistry*, 81,660-665.

Maeda, T., Kim, J. H., Ubukata, Y., & Morita, N. (2008). Analysis of volatile compounds in polished-graded wheat flours using headspace sorptive extraction. *European Food Research and Technology*, 227, 1233-1241.

Maeda, T., Kim, J. H., Ubukata, Y., & Morita, N. (2009). Analysis of volatile compounds in polished-graded wheat flour bread using headspace sorptive extraction. *European Food Research and Technology*, 228, 457-465.

Maeda, T., Maeda, N., & Morita, N. (2001). Effect of polished-graded hard-type wheat flour substitution for commonly milled hard-type wheat flour on the properties of dough and bread. *Journal of Applied Glycoscience*, 48, 27-36.

Maeda, T., Ohkura, M., & Morita, N. (1999). Characterization of wheat flour graded by polishing and its application to breadmaking. *Journal of Applied Glycoscience*, 46, 413-422.

McWilliams, M., & Mackey, A C. (1969). Wheat flavor components. *Journal of Food Science*, 34, 493-496.

Röcken, W. (1996). Applied aspects of sourdough fermentation. *Advances in Food Sciences*, 18, 212-216.

Röcken, W., & Voysey, P. A. (1995). Sourdough fermentation in bread making. *Journal of Applied Bacteriology-Symposium Supplement*, 79, 38-48.

Rounaud, O., & Martínez-Anaya, M. A. (1997). Relationships between biochemical and quality-related characteristics of breads, resulting from the interaction of flour, microbial starter and the type of process. *Food Research and Technology,* 204, 321-326.

Salinas, M. R., Zalacain, A., Pardo, F., & Alonso, G. L. (2004). Stir bar sorptive extraction applied to volatile constituents evolution during vitisvinifera ripening. *Journal of Agricultural and Food Chemistry,* 52, 4821-4827.

Salmenkallio-Marttila, M., Katina, K., &Autio, K. (2001). Effects of bran fermentation on quality and microstructure of high-fiber wheat bread. *Cereal Chemistry*, 78, 429-435.

Seitz, L. M., & Sauer, D. B. (1994). Detection of specific compounds that indicate off-odors in grains. *Cereal Foods World,* 39, 605.

Seitz, L. M., Ram, M. S., & Rengarajan, R. (1999). Volatiles obtained from whole and ground grain samples by supercritical carbon dioxide and direct helium purge methods:Observations on 2,3-butanediols and halogenated anisoles. *Journal of Agricultural and Food Chemistry*, 47, 1051-1061.

Wasowicz, E., & Kamiński, E. (1988).Volatile compounds of sound and musty wheat grains. *Chemie, Microbiologie, Technologie der Lebensmittel*, 11, 161-168.

Wehrle, K., & Arendt, E. K. (1998). Rheological changes in wheat sourdough during controlled and spontaneous fermentation. *Cereal Chemistry*, 75, 882-886.

In: Advances in Cereal and Pseudocereal ... ISBN: 978-1-62618-347-6
Editors: N. Morita, P.V. Hung et al.

Chapter 3

ADVANCES IN WAXY AND HIGH-AMYLOSE WHEAT RESEARCH

Pham Van Hung[1,*] and Naofumi Morita[2]
[1]School of Biotechnology, International University, National University in HoChiMinh City, HoChiMinh City, Vietnam
[2]Department of Food Packaging Technology, Toyo College of Food Technology, Mimami-Hanayashiki, Kawanishi, Hyogo, Japan

ABSTRACT

Waxy (amylose-free) and high-amylose wheats are new cultivars produced by genetic recombination. Biochemical and molecular techniques changed the starch structures of the genetic wheat, improving the unique characteristics and functionality of the starch. Waxy and high-amylose wheat flours have higher protein, lipid, ash and dietary fiber contents than non-waxy wheat flour. The large amount of dietary fiber is considered to be responsible for the characteristics of high-amylose wheat flour. However, doughs made from the waxy and high-amylose wheat flours exhibit less stability during mixing and inferior gluten matrix due to inferior protein quality and a high amount of dietary fiber. As a result, waxy and high-amylose wheat flours cannot be used solely for breadmaking due to the low loaf volume and the inferior appearance of the breadcrumb. However, the waxy wheat flour could be used to partly substitute for regular wheat flour to make soft, sticky and glutinous

* E-mail: pvhung@hcmiu.edu.vn.

breadcrumbs and retain the softness of breadcrumbs by reheating during storage. The waxy wheat may be substituted for up to 50% white wheat flour in whole wheat bread to improve the quality of bread. Likewise, high-amylose wheat flour could be used to substitute for regular wheat flour to increase the total amounts of dietary fiber and resistant starch in bread, increasing the beneficial physiological effects on human health. Thus, waxy and high-amylose wheats could be used for breadmaking to improve the texture and health benefits of the end-use products.

INTRODUCTION

Wheat, cereal plant of the genus *Triticum* of the family Gramineae (grass family), is a major food and an important commodity on the world grain market. The origin of the wheat plant is believed to be developed from a type of wild grass native to the arid lands of Asia Minor. The various forms of the genus *Triticum* were mainly grouped as einkorn (*T. monococcum*), emmer (*T. dicoccum*), and spelt (*T. spelta*). Bread wheat (*Triticum aestivum L.*) originated by hybridization from an emmer type and a wild species of grass (Shellenberger, 1971). Presently, wheat is produced in a wide range of climates and soil conditions, and is grown as far north as the Arctic Circle and as far south as the equator. In terms of global production, wheat is the second-largest cereal crop, tied with maize, and the third is rice. In 2011, the global wheat production was 691 million tons (FAO, 2011).

A wide variety of wheat cultivars have been developed with different qualities accompanying the development of genetic recombination. Waxy wheat (amylose-free, glutinous) and high-amylose wheat grains have been recently produced in Japan and elsewhere (Kiribuchi-Otobe et al., 1997; Nakamura et al., 1995; Yamamori et al., 2000; Yasui et al., 1997). The waxy and high-amylose wheats with different composition and functionality are believed to enhance the quality of wheat-based food products. Research on characteristics and application of waxy and high-amylose wheats has been conducted with the intent to have these cultivars becoming commercial products. The objective of this chapter is to summarize the results of the recent research on the characteristics and functionality of waxy and high-amylose wheats and their application for food processing.

Characteristics of Starches from Mutant Wheat Varieties

In plants, starch is synthesized in specialized organelles known as amyloplasts and consists of two kinds of glucose polymers, amylose and amylopectin. Two distinct types of starch synthases, isoforms of granule-bound starch synthase (GBSS, EC 2.4.1.21) and soluble starch synthases, are involved in the conversion of ADP-glucose to starch polymers (Buleon et al., 1998). GBSS I, also known as the waxy (*Wx*) protein, has a molecular weight of about 60 kDa and is responsible for the biosynthesis of the amylose fraction (Nelson and Rines, 1962), while amylopectin synthesis is more complicated with concerted activities of the soluble starch synthases together with branching and de-branching enzymes (Ball et al., 1996).

The waxy wheat was produced by using traditional hybridizations between the *Wx-D1* single null line BaiHuo and the *Wx-A1*/*Wx-B1* double null line "Kanto 107" (Nakamura et al., 1995), by treating seeds of the double null line Kanto 107 with a mutagen, ethyl methanesulfonate (Yasui et al., 1997), or by removing rapidly a low-amylose characteristic from a mutant line using the haploid breeding method (Kiribuchi-Otobe et al., 1997). In contrast, high-amylose wheat was produced by genetically modified SGP-1 null mutant, which enhances apparent amylose content and generates altered amylopectin (Yamamori et al., 2000). Since cv. Kanto 107 is known to be deficient in both Wx-A1 and Wx-B1 proteins, a mutation at the *Wx-D1* locus can also produce a waxy and low-amylose genotype without a substantial change of the genetic background of the parent. *Yasui* et al. (1997, 1998) reported that treating cv. Kanto 107 seeds with ethyl methanesulfonate produced two waxy endosperm mutant lines, K107Wx1 and K107Wx2, with non-functional *Wx-D1d.* K107Wx1 and K107Wx2 are the first waxy lines of hexaploid wheat induced by mutagen treatment for the mutation at the *Wx-D1a* gene in cv. Kanto 107 resulting in the lack of deposition of *Wx* proteins in the endosperm starch of the mutant lines (Yasui et al., 1997). The amylose content of the endosperm starch from both mutant lines was 0.9% and the agronomic characters of K107Wx1 and K107Wx2 were very similar to those of cv. Kanto 107 grown under field conditions. However, the waxy endosperm mutant lines had a recessive waxy gene with two other nonfunctional *Wx-A1* and *Wx-B1* genes, because the F1 from the cross between either mutants and the wild type, cv. Kanto 107, yielded seeds with a non-waxy endosperm. The endosperm of F2 seeds segregated into non-waxy and waxy with the ratio of 3:1, whereas the segregation ratio of non-waxy:waxy seeds of the test cross between waxy mutant lines and F1 was 1: 1 (Yasui et al., 1997). Another study on the cross between waxy mutants with the crossbred waxy lines (Wx-3 and Wx-4), which were derived from the cross between cv. Bai Huo and cv. Kanto 107, resulted in F1 seeds with a waxy endosperm. However, F2 seeds contained several non-waxy endosperms. The non-waxy F2 seeds were heterozygous, because the F3 seeds from the non-waxy F2 seeds segregated into non-waxy and waxy, which was consistent with the 3: 1 ratio (Yasui et al., 1998). Because of differences in the characteristics of starches of the mutant generations, further study on the backcrossing of mutant wheat lines is being done to improve the starch characteristics and wheat endosperm production. As a result, the near-isogenic wheat lines produced by the backcrossing of the K107Wx1 and K107Wx2 with cv. Kanto 107 are classified and the characteristics of their starches are investigated (Hung et al., 2008). Kanto 107*4/K107Wx1-nil and Kanto 107*4/K107Wx2-nil are near-isogenic lines produced by the recurrent backcross method using cv. Kanto 107 with functional *Wx-D1a* as the recurrent parent and waxy mutant lines induced from cv. Kanto 107, K107Wx1 or K107Wx2 (Yasui et al., 1997), with nonfunctional *Wx-D1d* as the non-recurrent parent. The pedigree of the near-isogenic lines is given in Figure 3-1. Heterozygous B1F1, B2F1 and B3F1 plants were crossed with the recurrent parent, cv. Kanto 107. Heterozygous B4F1 and B4F2 were self-pollinated to produce the next generation, and homozygous waxy and non-waxy B4F3 plants were self-pollinated to produce the seeds of the B4F4 generation.

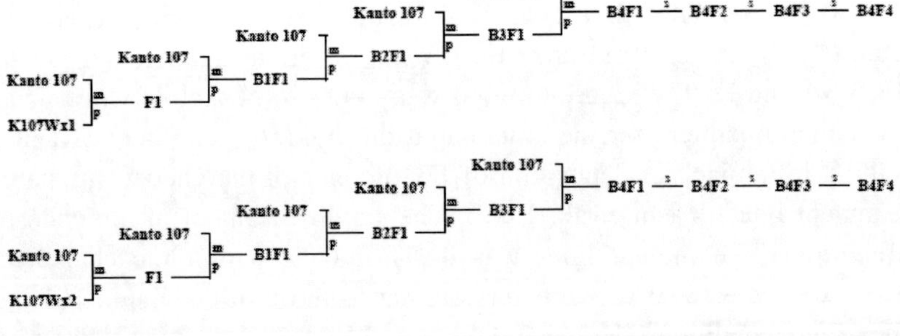

Figure 3-1. Pedigree of near-isogenic wheat lines (Adapted from Hung et al., 2008). Heterozygous B_1F_1, B_2F_1 and B_3F_1 plants were crossed with the recurrent parent, cv. Kanto 107. Heterozygous B_4F_1 and B_4F_2 were self-pollinated to produce the next generation, and homozygous waxy and non-waxy B_4F_3 plants were self-pollinated to produce the seeds of B_4F_4 generation. m, maternal parent; p, paternal parent; s, self-pollinated.

Finally, the homozygous B4F4 generations with waxy or non-waxy endosperm were grown in an experimental field at the National Agricultural Research Center for Western Region (Fukuyama, Japan) in 2004/2005 season under ordinary conditions and their seeds were used. Kanto 107, K107Wx1 and K107Wx2 were also grown in the same field and season and were used as controls. The starches of the mutant wheat lines were isolated and characterized for their structures and physicochemical properties. The amylose content of the mutant wheat starches are shown in Table 3-1. The two waxy mutant lines, K107Wx1 and K107Wx2 with non-functional *Wx-D1d* had 3.8% and 1.6% apparent amylose content, respectively as measured by the modified AACC method, whereas the reported amylose content of the mutant wheat was 0.9% measured by the concanavalin (ConA) method (Yasui et al., 1997). The two waxy near-isogenic lines, waxy Kanto 107*4/K107Wx1-nil and waxy Kanto 107*4/K107Wx2-nil, exhibited 3.0 and 2.7% apparent amylose content, respectively, which were not significantly different from their non-recurrent parent lines, K107Wx1 and K107Wx2. In contrast, the two non-waxy near-isogenic lines, non-waxy Kanto 107*4/K107Wx1-nil and non-waxy Kanto 107*4/K107Wx2-nil, had 25.4 and 23.2% apparent amylose content, respectively, which were slightly lower than the apparent amylose content of the recurrent parent line, cv. Kanto K107 (27.7 ± 2.0%). Blue values of the wheat starches were correlated with their amylose contents. The blue value of the non-waxy lines ranged in 0.242 – 0.283, whereas that of the waxy lines ranged in 0.049 – 0.060. Likewise, the wavelength of maximal starch-iodine absorption (λ_{max}) of the non-waxy wheat starch (568 – 575 nm) was significantly higher than that of the waxy lines (510 – 515 nm). The λ_{max} of starch-iodine complex of the waxy starch was also found to be lower than that of the non-waxy starch as previously reported by Hung et al. (2007a).

Table 3-1. Amylose contents of two sets of near-isogenic wheat starches[a] (Hung et al., 2008)

Cultivars	Amylose content (%)	Blue value	λ_{max} [nm]
Kanto 107	27.7 ± 2.0	0.283 ± 0.004	575 ± 0.5
K107Wx1	3.8 ± 0.5	0.060 ± 0.001	510 ± 0.5
K107Wx2	1.6 ± 0.5	0.049 ± 0.002	515 ± 1.0
Non-waxy Kanto 107*4/K107Wx1-nil	25.4 ± 1.5	0.249 ± 0.005	568 ± 1.4
Non-waxy Kanto 107*4/K107Wx2-nil	23.2 ± 1.0	0.242 ± 0.005	569 ± 2.0
Waxy Kanto 107*4/K107Wx1-nil	3.0 ± 1.0	0.055 ± 0.001	510 ± 0.5
Waxy Kanto 107*4/K107Wx2-nil	2.7 ± 0.5	0.056 ± 0.002	515 ± 1.0

[a] Values are the mean of triplicate measurements, n = 3.

The differences in amylose contents of the mutant wheat starches led to variation in the physicochemical properties of the wheat. Two non-recurrent parent lines, K107Wx1 and K107Wx2, had no significant difference in gelatinization temperatures. However, the transition enthalpy of K107Wx1 starch was significantly higher than that of K107Wx2 starch. The waxy and non-waxy near-isogenic lines, Kanto 107*4/K107Wx1-nil, which had similar gelatinization temperature and transition enthalpy, exhibited higher gelatinization temperatures than cv. Kanto 107 and lower transition enthalpy than K107Wx1. In contrast, the waxy and non-waxy near-isogenic lines, Kanto 107*4/K107Wx2-nil were significantly different in thermal characteristics of starches. The waxy Kanto 107*4/K107Wx2-nil had similar gelatinization temperatures but higher transition enthalpy than that of its non-recurrent parent, K107Wx2, whereas the non-waxy Kanto 107*4/K107Wx2-nil had similar gelatinization temperatures but lower transition enthalpy than that of its recurrent parent, cv. Kanto 107. Two waxy mutant lines, K107Wx1 and K107Wx2 with non-functional *Wx-D1d* exhibited A-type crystal like cv. Kanto 107, a Wx-A1 and Wx-B1 protein-deficient cultivar. The waxy and non-waxy near-isogenic lines, Kanto 107*4/K107Wx1-nil and Kanto 107*4/K107Wx2-nil, also exhibited the A-type crystal with the major peaks at 5.8, 5.2, 4.8 and 3.8 Å. Thus, the results show that the backcrossing did not change the crystalline structure of wheat starch.

Table 3-2. Pasting properties of two sets of near-isogenic wheat starches (8%, w/v)[a,b] (Hung et al., 2008)

Sample	T_g (°C)	T_p (°C)	V_p (BU)	V_f(BU)	BD (BU)	SB (BU)
Kanto 107	75b	93d	690c	990g	130a	300f
K107Wx1	63a	75c	1100e	590c	700g	- 510a
K107Wx2	63a	67a	590b	350a	370d	- 240d
Non-waxy Kanto 107*4/K107Wx1-nil	81c	93d	560a	880f	230b	320g
Non-waxy Kanto 107*4/K107Wx2-nil	81c	93d	550a	750e	260c	200e
Waxy Kanto 107*4/K107Wx1-nil	62a	69ab	1000d	550b	580e	- 450c
Waxy Kanto 107*4/K107Wx2-nil	62a	72bc	1100e	610d	610f	- 490b

[a] T_g, T_p: gelatinization and peak temperatures; V_p, V_f: peak and final viscosities; *BD*, breakdown; *SB*, setback.

[b] The same letter in the same column is not significantly different ($P < 0.05$), n = 2.

Pasting properties of the near-isogenic wheat starches determined by a viscoamylograph are shown in Table 3-2. Two waxy mutant lines, K107Wx1 and K107Wx2, had lower gelatinization and peak viscosity temperatures than cv. Kanto 107. K107Wx1 and K107Wx2 also had lower final viscosities but higher breakdown than did cv. Kanto 107. However, the peak viscosity of K107Wx1 was significantly higher than that of K107Wx2 and cv. Kanto 107, whereas the peak viscosity of K107Wx2 was significantly lower than that of cv. Kanto 107. The waxy near-isogenic lines also had lower gelatinization and peak temperatures, higher peak viscosities and breakdown and lower final viscosity than those of the non-waxy near-isogenic lines. In addition, the non-waxy near-isogenic lines exhibited lower peak and final viscosities and higher breakdown than did the recurrent parent line, cv. Kanto 107. Thus, the near-isogenic lines showed the typical pasting properties of the waxy and non-waxy starches, which was not significantly different from their parent lines.

FLOUR AND DOUGH PROPERTIES OF WAXY AND HIGH-AMYLOSE WHEATS

Wheat flour, a proportion of 60% by weight of whole wheat grain recovered after removal germ and bran fractions by modern milling techniques, is the major ingredient for wheat-based product processing. The quality of wheat flour depends on the quantity and characteristics of its components, which differ between the various wheat cultivars. The general composition analysis of waxy and high-amylose flours is summarized in Table 3-3. Waxy wheat flour (WWF), selected from the F2 progeny of CS (Kanto 107 7A)/CS (Kanto 107 4A) // CS (Bai Huo 7D) and backcrossed twice to the normal CS (Chinese Spring) (Miura et al., 1999), had considerably more moisture, wet gluten, protein and dietary fiber contents than non-waxy wheat flour. High-amylose wheat flour (HAWF), selected from the F5 and F6 progeny with no SGP-1 of Kanto 79/Turkey 116 F2 //Chousen 57 (Yamamori et al., 2000), showed the significantly highest values of protein, ash, lipid and dietary fiber among all the tested samples. The large amount of dietary fiber was considered to be responsible for the characteristics of HAWF (Morita et al., 2002a). The sodium content was not distinctly different among the samples, but it was slightly lower in HAWF than in the other samples. The water activity of all samples was about 0.98, but the water activity of CSF was significantly higher than those of HAWF and WWF.

Table 3-3. Summary of general characteristics of waxy and high-amylose wheat flours[a,b,c] (Morita et al., 2002a)

Sample	General compositions (%)										
	AM	AP	Water	Wet gluten	Protein	Ash	Lipid	DF	*L*	Na (mg/100g)	a_w
CSF	24.9b	75.1b	12.5b	46.9b	14.9a	0.51a	0.8a	2.2a	95.0b	0.65a	0.989b
HAWF	37.5c	62.5a	11.3a	43.4a	17.5c	1.08c	1.4b	6.7c	90.8a	0.65a	0.984a
WWF	2.3a	97.7c	13.0c	51.3c	16.0b	0.54b	1.0a	3.3b	95.2b	0.70a	0.983a

[a] CSF, test milled soft-type wheat from Chinese Spring; HAWF, high-amylose wheat flour; WWF, waxy wheat flour.

[b] AM, amylose; AP, amylopectin; DF, dietary fiber; *L*, lightness value; a_w, water activity.

[c] Values followed by the same letter in the same column are not significantly different ($P < 0.05$).

The pasting properties of the high-amylose and waxy wheat flours are shown in Table 3-4. The maximum viscosity temperature of WWF paste was significantly lower than that of CSF paste, whereas the maximum viscosity of WWF paste was higher than that of CSF paste. This indicates that the starch in WWF gelatinizes quickly and has a high viscosity. Araki et al. (2000) reported that the pasting properties of waxy starch were affected by the null mutation at the three *Wx* loci. The effect of the *Wx-B1b* allele, giving a higher peak and breakdown viscosity, was distinct, while the presence or absence of the *Wx-A1* protein gave some variations in peak and breakdown viscosities and the effects of the *Wx-A1b* were much smaller than those of the *Wx-B1b* allele.

Table 3-4. Summary of pasting properties of waxy and high-amylose wheat flours[a,b,c] (Morita et al., 2002a)

Sample	GT	MVT	MV
CSF	57.7c	90.3b	481b
HAWF	45.3a	93.0c	308a
WWF	57.0b	66.1a	636c

[a] Gelatinization temperature (°C); MVT, maximum viscosity temperature (°C); MV, maximum viscosity (BU).

[b] CSF, test-milled soft-type wheat from Chinese Spring; HAWF, high-amylose wheat flour; WWF, waxy wheat flour.

[c] Values followed by the same letter in the same column are not significantly different ($P < 0.05$).

In addition, the null *Wx-D1b* allele produced a higher viscosity than the wide-type *Wx-D1a*. The gelatinizing temperature of HAWF starch was distinctly lower than that of CSF and WWF starches. In contrast, HAWF starch had a significantly higher maximum viscosity temperature than the other starches, while its maximum viscosity was significantly lower. Thus, the different amylose/amylopectin ratios of starches significantly affect the pasting properties of flour.

Dough - a complex mixture of ingredients including flour, water, yeast and salt - is one of the principal intermediate stages in the transformation of wheat flour to the end-use products such as noodles, breads and cookies. The dough derives its properties from the constituent components of flour, water and air. The role of the constituents of wheat flour in the dough was recently summarized in several reviews (Dobraszczyk and Morgenstern, 2003; Goesaert et al., 2005), in which the quantity and quality of gluten proteins were found to largely determine dough mixing requirement and rheological properties of the optimally mixed dough. The strong wheat flours with high protein content and desirable gluten quality made the good dough properties, which had high water absorption and stability time during mixing in a farinograph, and had high resistance and extensibility to stretching as determined by an extensograph (Hung et al., 2005a). In contrast, the doughs made with soft wheat flours containing low amount of protein and inferior gluten quality exhibited low water absorption and less stability during mixing, and less extensibility as compared with the doughs made from strong wheat flours (Hung et al., 2004). Starch is present in the dough in the native state and absorbs up to about 46% of water during dough preparation (Goesaert et al., 2005). However, the role of starch and its amylose/amylopectin ratio in the dough is still ambiguous. In addition, the minor groups such as lipid and non-starch polysaccharides also influence on dough properties (Goesaert et al., 2005). The doughs made from WWF and HAWF had significantly higher water absorption than the non-waxy wheat flour, which was due to high amount of protein and dietary fiber (Morita et al., 2002a). However, the dough made from WWF and HAWF exhibited lower stability during mixing similar to the soft wheat flours (Morita et al., 2002a; Hung et al., 2005b). The waxy wheat made the dough stickier and less firm than the non-waxy wheat, while the high-amylose wheat flour made a firmer and more viscous dough than the non-waxy wheat flour, but produced a softer and less elastic dough as compared to the hard wheat flour (Morita et al., 2002a; Hung et al., 2005b). The appearance of doughs prepared from WWF and HAWF, observed using scanning electron microscope, are given in Figure 3-2.

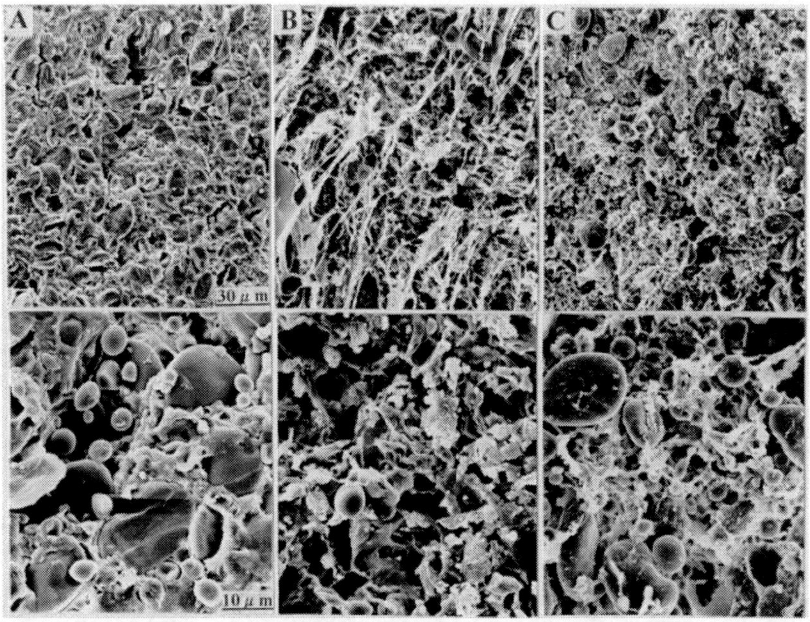

Figure 3-2. Scanning electron microscopy of dough from three wheat flours. A, Chinese Spring flour (CSF); B, high-amylose wheat flour (HAWF); C, waxy wheat flour (WWF). (Adapted from Morita et al., 2002a).

The HAWF dough (Figure 3-2B) had a rougher structure than the doughs of CSF (Figure 3-2A) or WWF (Figure 3-2C). The gluten sheet in the HAWF dough was irregular and discontinuous, and did not cover some starch granules. This result may be due to the presence of a larger amount of dietary fiber in HAWF. The gluten of WWF was not as evenly dispersed as it was in CSF dough and did not cover the starch granules as uniformly or entirely as it did in CSF dough. Thus, the inferior dough properties of the waxy and high-amylose wheat flours may affect the quality of end-use products, especially breadmaking quality.

Application of Waxy and High-Amylose Wheat for Breadmaking

Bread is an unstable, elastic and solid foam. The solid part of bread contains a continuous phase composed in part of an elastic network of cross-linked gluten molecules and in part of leached starch polymer molecules, primarily amylose, both uncomplexed and complexed with polar lipid molecules, and a discontinuous phase of entrapped, gelatinized, swollen, deformed wheat starch granules (Gray and BeMiller, 2003). Therefore, starch plays an important role in the textural characteristics of bread. With cooling and aging of bread, rearrangements in the starch fractions lead to large changes in the texture of bread including in gelation and crystallization. This starch retrogradation is believed to be the major cause of breadcrumb firming, commonly referred to as bread staling (Hug-Iten et al., 1999). However, whether the fraction of starch that contributes to bread firming is amylose or amylopectin has been also debated (Gray and BeMiller, 2003). Using waxy starch to change amylose:amylopectin ratio of in flour for studying the bread staling has been widely researched (Morita et al., 2002b; Lee et al., 2001; Hayakawa et al., 2004; Bhattacharya et al., 2002; Ghiasi et al., 1984). These studies reported that incorporation of waxy wheat starch resulted in retaining more moisture in bread-crumbs and retardation the staling and extension the shelf-life of breads. The firmness of breadcrumbs made from 20 and 40% of WWF substitutions compared with those from 20 and 40% of CSF or the control are given in Figure 3-3. The breadcrumbs made from 20 and 40% of WWF substitutions had significantly lower firmness than did the control after baking and during storage for 7 days, whereas the substitutions with 20 and 40% of CSF made the breadcrumbs firmer than the control.

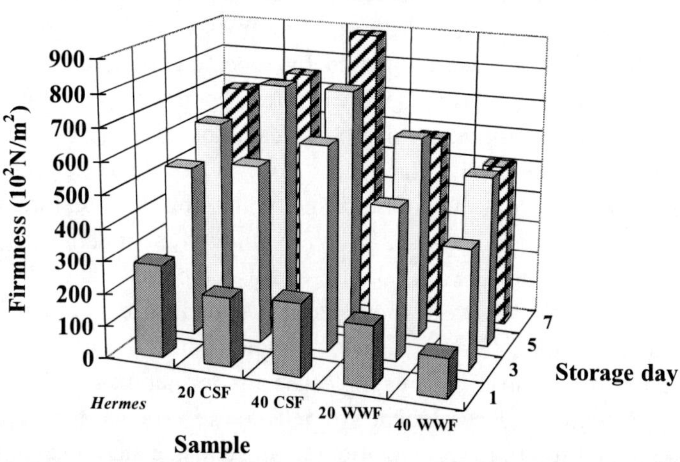

Figure 3-3. Effect of substitution of waxy wheat flour on firmness of breadcrumbs during storage. Hermes, foreign hard wheat flour; 20 CSF, CSF (20%)+Hermes (80%); 40 CSF, CSF (40%)+Hermes (60%); 20 WWF, WWF (20%)+Hermes (80%); 40 WWF, WWF (40%)+Hermes (60%) (Morita et al., 2002b).

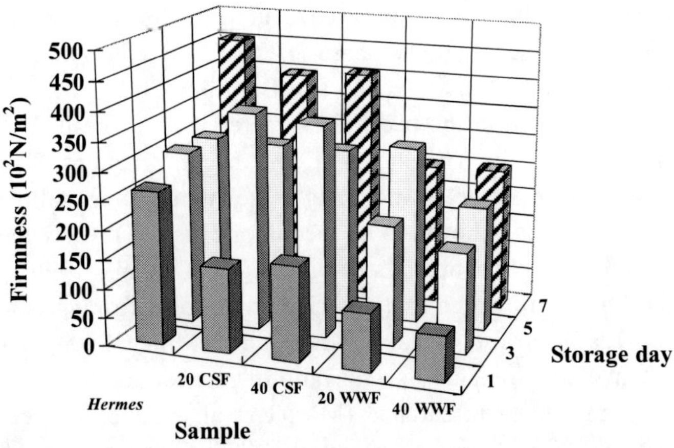

Figure 3-4. Effect of substitution of waxy wheat flour on refreshing of breadcrumbs by reheating. Hermes, foreign hard wheat flour; 20 CSF, CSF (20%)+Hermes (80%); 40 CSF, CSF (40%)+Hermes (60%); 20 WWF, WWF (20%)+Hermes (80%); 40 WWF, WWF (40%)+Hermes (60%) (Morita et al., 2002b).

Thus, the presence of WWF softened the breadcrumbs and this tendency increased along with the increase in the amount of CSF substitution. Thus, the increase in the percentage of amylopectin by substitution of WWF retarded the staleness of breadcrumbs. The substitutions of WWF not only improved the softness of breadcrumbs but also retained the softness of breadcrumbs after reheating (Figure 3-4). Although the breadcrumbs made from the control or any of the substitutions were softened by reheating regardless amounts of substitutions or storage days, the breadcrumbs containing WWF refreshed their staleness by reheating more distinctly than those with CSF. Thus, the use of WWF to substitute for regular wheat flour in breadmaking improves the quality of bread to give it the softness, sticky and glutinous texture, preferred by fastidious customers. Likewise, HAWF's unique functionality may be used to improve the nutritional value of the end-use products. Dietary fiber and resistant starch (RS) are well known to have healthy benefits for humans (Brown et al., 2001; Slavin, 2002; Topping and Clifton, 2001). The retrogradation of amylose is a major cause for formation of RS. Therefore, starches with high amylose content are believed to produce high amounts of RS in baking and food processing. High-amylose maize starch is the preferred starting material for producing high-RS ingredient and many publications have reported the formation of RS from high-amylose maize starch (Thompson, 2000). Therefore, HAWF with high amounts of dietary fiber and amylose is a good means of increasing the dietary fiber and resistant starch content in end-use food products. Although HAWF is found to have inferior dough and breadmaking qualities, it can be used to substitute for regular wheat flours to increase its functionality. The results of qualities of breads with 10, 30 and 50% of HAWF substitutions are shown in Table 3-5. Loaves of breads made from 10 and 30% of HAWF did not significantly differ from that of the control, whereas the substitution with 50% of HAWF made the loaf volume of bread significantly lower. This result is due to the inferior gluten quality and high amount of dietary fiber of HAWF as previously reported by Morita et al. (2002a). After baking, firmness of breadcrumbs increased along with increase in the amount of HAWF substitutions.

During storage, the firmness of breadcrumb with 10% of HAWF substitutions was higher than that of the control, whereas the firmness of breads with 30 and 50% substitutions did not significantly differ from that of the control. These results may be due to the high moisture content of breadcrumbs with 30 and 50% of HAWF substitutions after 3 and 5 days of storage. The substitution with HAWF increased the dietary fiber and resistant starch content in bread as shown in Figure 3-5. After baking, the amount of total dietary fiber and resistant starch in bread with 10, 30 and 50% of HAWF substitutions were 120, 160 and 177%, respectively, as compared to those in the control.

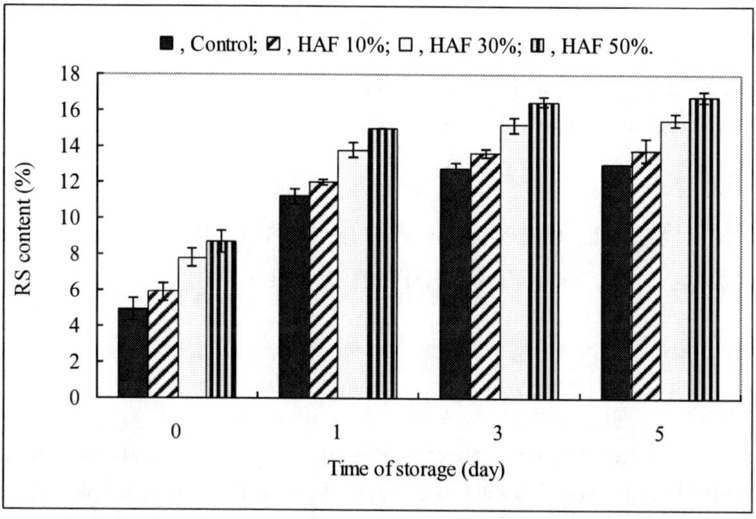

Figure 3-5. Total dietary fiber and resistant starch contents (%, db) in breads during storage with high-amylose wheat flour (HAF) substitutions (Adapted from Hung et al., 2005b).

Table 3-5. Quality of breads with high amylose wheat flour (HAF) substitutions[a,b] (Hung et al., 2005b)

Sample	Specific volume (g/cm^3)	Firmness (10^2 N/m^2) during storage (day)			
		0	1	3	5
Control	3.5b	36.2a	85.8a	143.2a	169.0b
HAF 10%	3.3b	44.8b	101.4b	167.5b	194.5c
HAF 30%	3.2b	49.3c	90.7a	140.2a	173.8b
HAF 50%	3.0a	61.3d	92.0a	130.7a	152.0a

[a] Control, Cameria 100%.

[b] The same letter in the same column is not significantly different ($P < 0.05$), n = 3.

The amount of RS increased rapidly after 1 day of storage and then rose gradually for next days of storage. The presence of high-amylose starch produced higher amount of RS than regular wheat starch. With 30 and 50% of HAWF substitutions, the total dietary fiber and RS contents in bread after 5 days of storage were 15.5 and 16.8% (db), respectively, as compared with 13.0% (db) of RS in bread from regular wheat flour. The significant increase of RS in bread after baking and the first day of storage was due to amylose retrogradation which mainly contributed to formation of RS in bread and performed rapidly in a few hours (Eerlingen et al., 1993; Sievert and Pomeranz, 1989). Thus, the increase in amount of RS in bread during storage illustrated that the retrogradation of amylose after gelatinization is major cause of formation of RS. As a result, high-amylose wheat flour might be used to substitute for up to 50% of normal wheat flour to make bread with acceptable bread quality and significantly high amount of RS.

Application of Whole Waxy Wheat for Breadmaking

The consumption of whole grains, which are rich in dietary fiber, resistant starch, vitamins, minerals and microconstituents, is reported to have many physiological benefits related to "western diseases" such as coronary heart disease, colon cancer and diabetes. However, whole grain foods are not as attractive to consumers because the higher bran and germ in whole grain flour reduces the quality and sensory value of the end-use products. In breadmaking, the presence of bran and germ deteriorates the rheological properties of dough, lowers loaf volume of bread, increases crumb firmness, darkens crumb appearance and gives different flavor profiles of whole wheat breads from those of white flour (Chang and Chambers, 1992; Lai et al., 1989; Ozboy and Koksel, 1997; Wang et al., 2002). This is in contrast to the white wheat flour, which is recovered by removing all bran and germ had the high functional properties. Breads made from the white wheat flour are more preferred by consumers than those from whole grain flours. The quality of whole grain breads may be improved by addition of various enzymes and additives. However, the inferior quality of whole grain breads is still a significant issue and scientific research has been focused on improving the quality of whole grain foods.

The qualities of bread made from whole waxy wheat, white waxy flour and whole regular wheat flour are given in Table 3-6 (Hung et al., 2007c). Although whole waxy wheat flour had lower protein content and higher amount of dietary fiber, the bread had similar specific volume to the regular whole wheat flour. However, the bread made from whole waxy wheat had a significantly lower specific volume than that from the white waxy flour. Thus, a large amount of dietary fiber in whole waxy wheat reduced the loaf volume of bread as compared to the white waxy flour. Firmness of breadcrumbs from the whole waxy wheat was significantly softer than those from the whole regular wheat after baking and during 3 days of storage (Table 3-6). However, the ratio of retrogradation of breadcrumbs from the whole waxy wheat after 1 day of storage was larger than that of breadcrumbs from the whole regular wheat. Therefore, the breadcrumbs from the whole waxy wheat after 5 days of storage were firmer than that from the whole regular wheat. These results indicate that the staleness of breadcrumbs was firstly caused by the retrogradation of amylose fraction of starch and then caused by the crystallization of amylopectin fraction, which resulted in the larger ratio of the firmness of breadcrumbs.

Table 3-6. Quality of bread made from whole regular, whole waxy and white waxy flour[a,b] (Hung et al., 2007c)

Sample	Specific volume (cm^3/g)[a]	Firmness (10^2 Nm^{-2})[b]			
		0 day	1 day	3 days	5 days
Whole regular wheat	1.8 ± 0.2	133.6b	179.8c	256.2c	305.9b
Whole waxy wheat	1.8 ± 0.2	31.0a	48.5b	210.2b	333.5c
White waxy flour	4.1 ± 0.2	nd	36.8a	116.3a	137.2a

[a] The value shows the mean of triplicate data.

[b] The same letter in the same column is not significantly different ($P < 0.05$), n = 3.

Because the quality of bread made from 100% of whole grain flour was very poor, the amount of whole wheat flour used was less than 50% of the substituted flour in the formulas (Hung et al., 2007b). The dough made from WWF had significantly higher water absorption than CWF. The increasing amount of WWF substitutions also significantly increased their amounts of water absorptions. The high amount of water absorption was caused by the existence of high amount of protein and dietary fiber in the doughs. The high amount of water absorption may affect the fermentation of dough and bread staling during storage. It also causes difficulty for bakers to prepare and bake the dough. Furthermore, the dough made from WWF was weaker and less stable during mixing than CWF. The substitution with WWF decreased the development and stability times of doughs during mixing. The appearances of doughs made from WWF and the substituted flours exhibited that the dietary fiber combined with protein to form a discontinuous and irregular matrix around the starch granules, whereas the dough made from CWF had an extensible gluten matrix which covered all starch granules (Figure 3-6). Thus, the existence of dietary fiber diluted the protein and interfered with optimal gluten matrix formation during dough mixing. The dilution also affects the formation of an elastic network of cross-linked gluten molecules during baking, resulting in easy disruption of gluten network and low loaf volume of bread.

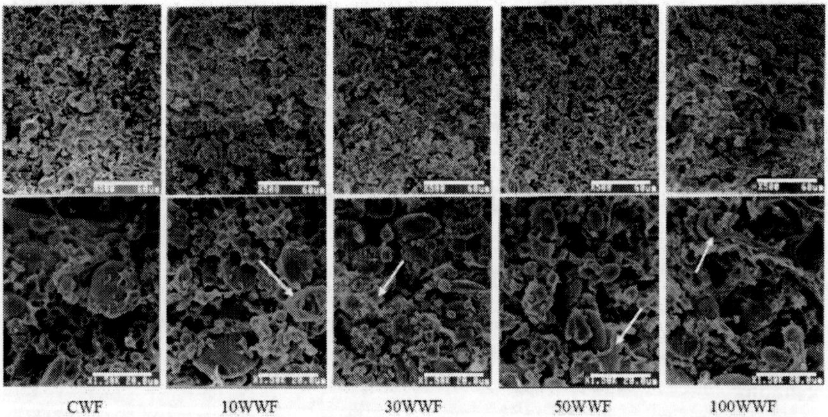

Figure 3-6. Scanning electron microscopy (SEM) of doughs made from whole waxy wheat flour (WWF) and its substitutions (Adapted from Hung et al., 2007b). CWF, Commercial white wheat flour; 10WWF, 30WWF, 50WWF and 100WWF are 10%, 30%, 50% and 100% of whole waxy flour substitutions. Arrows show dietary fiber appeared in doughs.

Table 3-7. Quality of bread made from whole waxy wheat flour (WWF) and its substitutions[a,b] (Hung et al., 2007b)

Sample	Specific volume (cm^3/g)	Firmness (10^2 Nm^{-2})		
		0 day	1 day	3 days
CWF	3.3 ± 0.2	39.0c	86.6d	153.2a
10WWF	3.0 ± 0.2	43.5d	91.8d	172.3b
30WWF	2.9 ± 0.2	40.3cd	79.0c	173.0b
50WWF	2.7 ± 0.2	34.8b	58.3b	176.2b
100WWF	2.2 ± 0.2	27.6a	36.8a	150.2a

[a] Abbreviations are the same as in Figure 3-6.
[b] The same letter in the same column is not significantly different (P <0.05), n = 3.

Breads made from whole waxy wheat flour as well as from 10%, 30% and 50% of WWF substitutions had lower specific volumes as compared to bread made from CWF or the substituted flours. The increasing amount of substitution also decreased the specific volume of bread (Table 3-7). Firmness of breadcrumbs made from WWF had significantly lower than those from CWF on the day of baking as well as during storage for three days. The 30% and 50% WWF substitutions, which absorbed high amount of water during dough preparation, also had high moisture content in breadcrumbs resulting in lower firmness on the day of baking and after one day of storage. Thus, the superior property of waxy wheat on retardation of bread staling is a significant benefit for application of whole waxy wheat in breadmaking.

The several studies reported that the addition of enzymes improved the bread specific volume and crumb firmness of breads made from white wheat flour (Courtin et al., 2001; Harada et al., 2005; Maeda et al., 2003; Shah et al., 2006). In order to determine effects of different enzymes on bread quality made from the whole waxy wheat, the cellulase, pentosanase and α-amylase with the same amount of enzymes (5 U/g) were used to make whole waxy wheat breads (Hung et al., 2007c). The addition of cellulase slightly reduced specific volume of bread, whereas pentosanase improved the loaf volume (Figure 3-7).

However, the addition of pentosanase increased firmness of the bread, while the addition of cellulase made the bread softer (Figure 3-8). The addition of α-amylase did not affect bread quality both in specific volume and firmness of bread (Figs. 3-7 and 3-8). Thus, the presence of the hydrolysate from the addition of cellulase and pentosanase to whole waxy wheat had different impact on bread quality. Maeda and Morita (2003) reported that a relatively larger amount of water-insoluble pentosans as compared to water-soluble pentosans is needed to obtain good baking quality. As a result, an optimum level of cellulase or pentosanase could improve the quality of bread made from whole waxy wheat. Harada et al. (2005) also reported that an optimum level of enzymes increased loaf volume and bread score and reduced crumb firmness tested with a straight-grade flour of Canadian red spring wheat. However, at high levels of addition, loaf volume and bread score decreased and crumb firmness increased as compared to the addition of optimum level. For the whole wheat bread, the addition of xylanase at ratio of 12 U/g flour was reported to improve specific volume of bread by 56% and reduce the firmness of bread by 77% (Courtin et al., 2001).

Although breads made from whole waxy wheat were significantly softer, their loaf volumes were still small. However, the breadcrumb from whole waxy wheat had big gas cells and opened crumbs as compared to that from whole regular wheat which had small gas cell and tightly structure of crumbs (Figure 3-9). In addition, the breadcrumbs from whole waxy wheat were sticky and glutinous. The addition of enzymes did not change the gas cell structure of whole waxy wheat breadcrumbs.

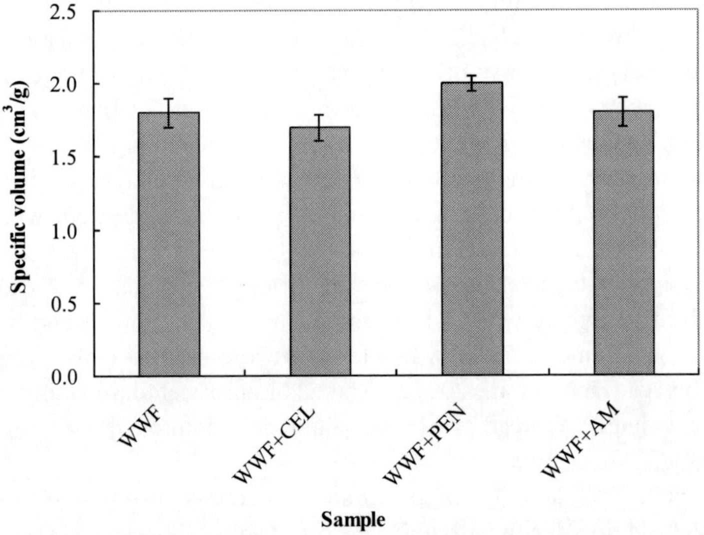

Figure 3-7. Effects of additional enzymes on specific volumes of breads baked from whole waxy wheat flour (adapted from Hung et al., 2007c). WWF, whole waxy wheat; WWF+CEL, whole waxy wheat added cellulase; WWF+PEN, whole waxy wheat added pentosanase; WWF+AM, whole waxy wheat added α-amylase.

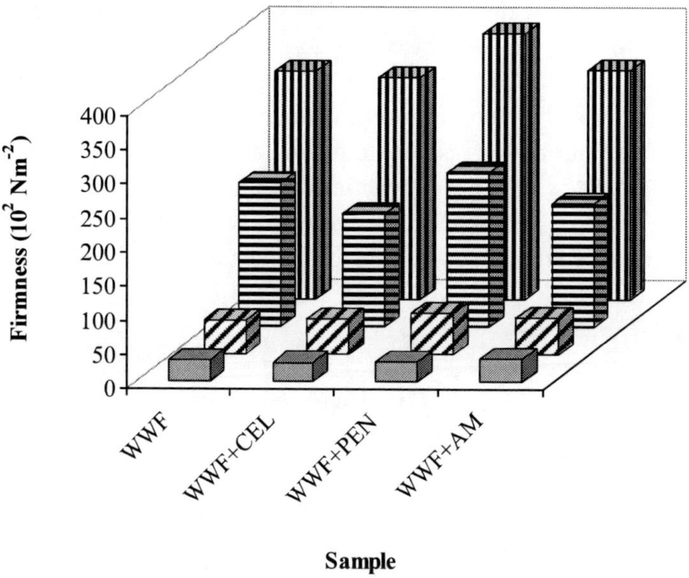

Figure 3-8. Effects of additional enzymes on firmness of breadcrumbs baked from whole waxy wheat flour (Adapted from Hung et al., 2007c). Abbreviations are the same as in Figure 3-7. ■, 0 day; ▨, 1 day; ▤, 3 days; ▥, 5 days.

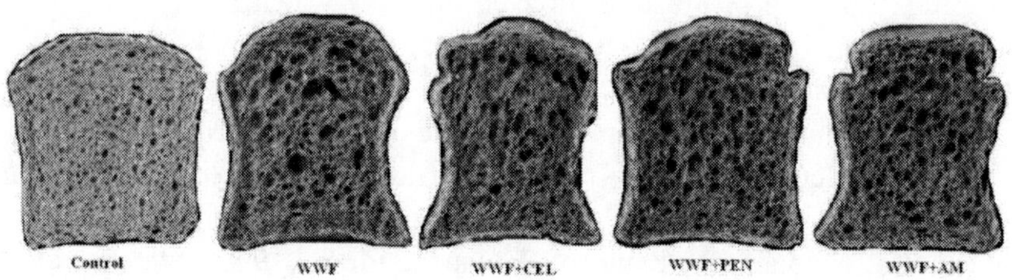

Figure 3-9. Cross-sectional views of breads baked from whole waxy wheat flour with additional enzymes (Adapted from Hung et al., 2007c). Abbreviations are the same as in Figure 3-7.

CONCLUSION

Waxy and high-amylose wheat cultivars are the new cultivars produced by genetic recombination. The structures of starches in wheat endosperm have been changed based on the biochemical and molecular techniques, which improves the unique characteristic and functionality of starch. However, waxy and high-amylose wheat flours cannot be used alone for breadmaking due to the low quality of doughs and breads. The waxy wheat flour could be used to partly substitute for regular wheat flour to make the soft, sticky and glutinous breadcrumbs and retain the softness of breadcrumbs by reheating during storage. Likewise, high-amylose wheat flour could be used to substitute for regular wheat flour to increase the total amounts of dietary fiber and resistant starch in breads which have beneficially physiological effects on human health. In addition, the superior property of waxy wheat on retardation of bread staling is a significant benefit for application of whole waxy wheat in breadmaking.

REFERENCES

Araki, E., Miura, H., and Sawada, S. (2000). Differential effects of the null alleles at the three Wx loci on the starch-pasting properties of wheat. *Theoretical and Applied Genetics, 100,* 1113-1120.

Ball, S., Guan, H.-P., James, M., Myers, A., Keeling, P., Mouille, G., Buleon, A., Colonna, P., and Preiss J. (1996). From glycogen to amylopectin: a model for the biogenesis of the plant starch granule. *Cell, 86*, 349-352.

Bhattacharya, M., Erazo-Castrejon, S., Doehlert, D. C., and McMullen, M. S. (2002). Staling of bread as affected by waxy wheat flour blends. *Cereal Chemistry, 79,* 178-182.

Brown, I. J., McNaught, K. J., Andrew, D., and Morita, T. (2001). Resistant starch: Plant breeding, application, development, and commercial use. In B. V. McCleary, and L. Prosky (Eds.), *Advanced Dietary Fiber Technology* (pp. 401-412), Iowa State University Press, Ames, IA.

Buleon, A., Colonna, P., Planchot, V., and Ball, S. (1998). Starch granules: structure and biosynthesis. *International Journal of Biological Macromolecules, 23,* 85-112.

Chang, C.-Y., and Chambers, E. (1992). Flavor characteristics of breads made from hard red winter wheat and hard white winter wheat. *Cereal Chemistry, 69*, 556-559.

Courtin, C. W., Gelders, G. G., and Delcour, J. A. (2001). Use of two endoxylanases with different substrate selectivity for understanding arabinoxylan functionality in wheat flour breadmaking. *Cereal Chemistry, 78*, 564–571.

Dobraszczyk, B. J., and Morgenstern, M. P. (2003). Rheology and the breadmaking process. *Journal of Cereal Science, 38,* 229-245.

Eerlingen, R. C., Deceuninck, M., and Delcour, J. A. (1993). Enzyme-resistant starch. II. Influence of amylose chain length on resistant starch formation. *Cereal Chemistry, 70*, 345-350.

Food and Agriculture Organization of the United Nations (FAO) (2011). Food Outlook – No. 1. *Food Outlook (GIEWS) - ESC/Reports*.

Ghiasi, K., Hoseney, R. C., Zeleznak, K., and Rogers, D. E. (1984). Effect of waxy barley starch and reheating on firmness of bread crumb. *Cereal Chemistry, 61,* 281-285.

Goesaert, H., Brijs, K., Veraverbeke, W. S., Courtin, C. M., Gebruers, K, and Delcour, J. A. (2005). Wheat flour constituents: how they impact bread quality, and how to impact their functionality. *Trends in Food Science and Technology, 16,* 12-30.

Gray, J. A., and BeMiller, J. N. (2003). Bread staling: molecular basis and control. *Comprehensive Reviews in Food Science and Food Safety, 2,*1-21.

Harada, O., Lysenko, E. D., Edwards, N. M., and Preston, K. R. (2005). Effects of commercial hydrolytic enzymes additives on Japanesestyle sponge and dough bread properties and processing characteristics. *Cereal Chemistry, 82,* 314-320.

Hayakawa, K., Tanaka, K., Nakamura, T., Endo, S., and Hoshino, T. (2004). End use quality of waxy wheat flour in various grain-based foods. *Cereal Chemistry, 81,* 666-672.

Hug-Iten, S., Handschin, S., Conde-Petit, B., and Escher, F. (1999). Changes in starch microstructure on baking and staling of wheat bread. *Food Science and Technology, 32,* 255-260.

Hung, P. V., Maeda, T., Yoshikawa, R., and Morita, N. (2004). Dough properties and baking quality of several domestic wheat flours as compared with commercial foreign wheat flour. *Food Science and Technology Research, 10,* 389-395.

Hung, P. V., Maeda, T., Yamauchi, H., and Morita, N. (2005a). Dough and breadmaking properties of various strong wheat grains cultivated in Japan. *Journal of Applied Glycoscience, 52,* 15-21.

Hung, P. V., Yamamori, M., and Morita, N. (2005b). Formation of enzyme-resistant starch in bread as affected by high-amylose wheat flour substitutions. *Cereal Chemistry, 82*, 690-694.

Hung, P. V., Maeda, T., and Morita, N. (2007a). Study on physicochemical characteristics of waxy and high-amylose wheat starches in comparison with normal wheat starch. *Starch/Staerke*, *59*, 125-131.

Hung, P. V., Maeda, T., and Morita, N. (2007b). Dough and bread qualities of flours with whole waxy wheat flour substitution. *Food Research International*, *40*, 273-279.

Hung, P. V., Maeda, T., Fujita, M., and Morita, N. (2007c). Dough properties and breadmaking qualities of whole waxy wheat flour and effects of additional enzymes. *Journal of the Science of Food and Agriculture*, *87*, 2538-2543.

Hung, P. V., Yasui, T., Maeda, T., and Morita, N. (2008). Physicochemical characteristics of starches of two sets of near-isogenic wheat lines with different amylose content. *Starch/Staerke*, *60*, 34-40.

Kiribuchi-Otobe, C., Nagamine, T., Yanagisawa, T., Ohnish, M., and Yamaguchi, I. (1997). Production of hexaploid wheats with waxy endosperm character. *Cereal Chemistry, 74,* 72-74.

Lai, C. S., Hoseney, R. C., and Davis, A. B. (1989). Effects of wheat bran in breadmaking. *Cereal Chemistry, 66*, 217-219.

Lee, M. R., Swanson, B. G., and Baik, B. K. (2001). Influence of amylose content on properties of wheat starch and breadmaking quality of starch and gluten blends. *Cereal Chem., 78,* 701-706.

Maeda, T., Hashimoto, T., Minoda, M., Tamagawa, S., and Morita N. (2003). Effects of mutant thermostable alpha-amylases on rheological properties of wheat dough and bread. *Cereal Chemistry 80*, 722–727.

Maeda, T., and Morita, N. (2003). Flour quality and pentosan prepared by polishing wheat grain on breadmaking. *Food Research International, 36*, 603–610.

Miura, H., Araki, E., and Tarui, S. (1999). Amylose synthesis capacity of the three Wx genes of wheat cv. Chinese Spring. *Euphytica, 108,* 91-95.

Morita, N., Maeda, T., Miyazaki, M., Yamamori, M., Miura, H., and Ohtsuka, I. (2002a). Dough and baking properties of high-amylose and waxy wheat flours. *Cereal Chemistry, 79,* 491-495.

Morita, N., Maeda, T., Miyazaki, M., Yamamori, M., Miura, H., and Ohtsuka, I. (2002b). Effect of substitution of waxy-wheat flour for common flour on dough and baking properties. *Food Science and Technology Research, 8,* 119-124.

Nakamura, T., Yamamori, M., Hirano, H., Hidaka, S., and Nagamine, T. (1995). Production of waxy (amylose free) wheats. *Molecular and General Genetics, 248,* 253-259.

Nelson, O. E., and Rines, H. W. (1962). The enzymatic deficiency in the waxy mutant of maize. *Biochemical and Biophysical Research Communications*, *9*, 297-300.

Ozboy, O., and Koksel, H. (1997). Unexpected strengthening effects of a coarse wheat bran on dough rheological properties and baking quality. *Journal of Cereal Science, 25*, 77-82.

Shah, A. R., Shah, R. K., and Madamwar, D. (2006). Improvement of the quality of whole wheat bread by supplementation of xylanase from *Aspergillus foetidus*. *Bioresource Technolonogy, 97*, 2047-2053.

Shellenberger, J. A. (1971): "Wheat, Chemistry and Technology", ed. Y. Pomeranz, pp. 1-18, American Association of Cereal Chemists, Inc., St. Paul, MN.

Sievert, D., and Pomeranz, Y. (1989). Enzyme-resistant starch I. Characterization and evaluation by enzymatic, thermoanalytical and microscopic methods. *Cereal Chemistry*, *66*, 342-347.

Slavin, J. L. (2002). Whole grains, dietary fiber, and resistant starch. In L. Marquart, J. L. Slavin, and R. G. Fulcher (Eds.), *Whole-grain foods in health and disease* (pp. 283-299). American Association of Cereal Chemists, Inc., St. Paul, MN.

Topping, D. L., and Clifton, P. M. (2001). Short-chain fatty acids and human colonic function: roles of resistant starch and nonstarch polysaccharides. *Physiological Reviews, 81,* 1031-1064.

Wang, J., Rosell, C. M., and de Barber, C. B. (2002). Effect of the addition of different fibres on wheat dough performance and bread quality. *Food Chemistry, 79*, 221-226.

Yamamori, M., Fujita, S., Hayakawa, K., Matsuki, J., and Yasui, T. (2000). Genetic elimination of starch granule protein, SGP-1, of wheat generates and altered starch with apparent high amylase. *Theoretical and Applied Genetics, 101,* 21-29.

Yasui, T., Matsuki, J., Sasaki, T., and Yamamori, M. (1997). Waxy endosperm mutants of bread wheat (*Triticum aestivum* L.) and their starch properties. *Breeding Science, 47,* 161-163.

Yasui, T. Sasaki, T., and Matsuki, J. (1998). Waxy bread wheat mutants, K107Wx1 and K107Wx2, have a new null allele on Wx-D1 locus. *Breeding Science, 48*, 405-407.

Section 2: Buckwheat-based Functional Food Development

N. Morita and P. V. Hung

Foreword

This section is dedicated to the memory Dr. Kazuyoshi Miyake, who was the former chairman of Miyake Flour Milling Company and was passed away in March, 2012. Some data of this section are from his PhD thesis.

In: Advances in Cereal and Pseudocereal ... ISBN: 978-1-62618-347-6
Editors: N. Morita, P.V. Hung et al.

Chapter 4

POLISHED-GRADING BUCKWHEAT FOR FUNCTIONAL FOODS

Pham Van Hung,[1] Kazuyoshi Miyake[2] and Naofumi Morita[3]

[1]School of Biotechnology, International University, National University in HoChiMinh City, HoChiMinh City, Vietnam
[2]Miyake Flour Milling Co., Momodani, Ikuno-ku, Osaka, Japan
[3]Department of Food Packaging Technology, Toyo College of Food Technology, Mimami-Hanayashiki, Kawanishi, Hyogo, Japan

ABSTRACT

A gradual milling method was recently used to mill whole buckwheat grains into various flour fractions. Sixteen fractions were obtained from the inner to the outer layers of the grain by this gradual milling method which were characterized and practically applied for noodle making in this chapter. Protein and ash content of flours increased in the order from inner to outer fractions. The phenolic and flavonoid contents of both free and bound phenolic extracts of buckwheat flour fractions significantly increased in the order from the fraction number 1 (phenolics less rich fraction) to the fraction number 16 (phenolics rich fraction). The phenolic compounds in buckwheat existed primarily in free form, whereas the flavonoids existed in insoluble bound forms, bound to cell wall materials. The optimum cooking time and the whiteness of uncooked and cooked noodles substituted with 40% of the fractions decreased from the inner to outer fractions, whereas the noodles made from the inner fractions were harder and less elastic than those from the outer fractions. As a result, the different buckwheat fractions contributed to different quality of soba noodles. However, the appropriate fractions could be used for noodle making to improve the quality of noodles depending on nutritional demands.

INTRODUCTION

Buckwheat (*Fagopyrum esculentum* Moench), a pseudocereal, is an alternative crop belonging to the *Polygonaceae* family and is usually grouped with cereals because of similarity in cultivation and utilization though it is not cereal grain. There are three major buckwheat species, common (*F. esculentum* Moench), tartary (*F. tataricum* Gaertner) and perennial (*F. cymosum* L.), of which common buckwheat is the most commonly grown species (Marshall and Pomeranz, 1982; Mazza and Oomah, 2005). In addition, Japanese common buckwheat is classified into two groups according to harvest season (summer and autumn) (Yoshimoto et al., 2004). Buckwheat was originally cultivated in southwest China and the Himalayan region (Ohnishi, 1998).

Presently buckwheat is grown as a traditional crop in China, Japan, Russia, Canada, USA, Australia, Central and Eastern Europe, etc. In Eastern Europe, buckwheat is a basic food item in porridges and soups. In North America it is marketed primarily in pancake mixes, which may contain buckwheat flour mixed with wheat, maize, rice, or oat flour, plus a leavening agent (Joshi and Rana, 1995). In Japan, buckwheat is marketed primarily as flour for manufacturing a variety of noodles (soba) (Udesky, 1988). Recently there has been renewed interest in buckwheat as an alternative crop for organic cultivation or as a functionally healthy food because of its low-calorie, high-fat content of unsaturated fatty acids, reasonable amino acid composition, high biological value of protein and potential antioxidative activity (Franschi et al. 1994; Hung et al. , 2007; Hung and Morita, 2008; Watanabe et al., 1997; Wijngaard and Arendt, 2006).

Buckwheat seed, known as achene, has a hard outer hull, which has a hard fibrous structure and surrounds the seed coat, endosperm and embryo tightly (Figure 4-1). The embryo in a buckwheat seed is located in the center of the endosperm and the endosperm contains large amounts of protein, starch and vitamins.

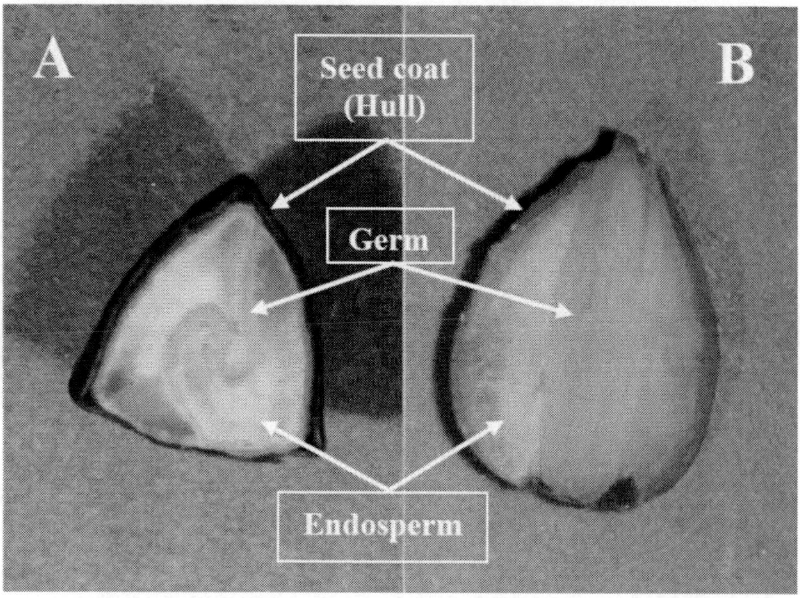

Figure 4-1. Image of Buckwheat grain. A, cross section; B, vertical section.

The protein of buckwheat consists of well-balanced amino acids with a high biological value (Pomeranz, 1983) and is an excellent supplement for cereal grains (Joshi and Rana, 1995). In addition, buckwheat grains have been well known as a plant source of rutin, quercetin, kaempferol-3-rutinoside, and a trace quantity of a flavonol triglycoside (Holasova et al., 2002). Like wheat, buckwheat bran and germ are rich in dietary fiber, minerals and phytochemicals which play important roles in human nutrition and health. Significant health benefits are derived from bran and germ such as dietary fiber or phenolic acids. The consumption of dietary fiber has been shown to reduce the risk of heart disease and to prevent colorectal cancer, and metabolic and inflammatory bowel diseases such as diabetes and diverticulitis (Craig et al., 1998; Topping and Clifton, 2001), while phenolic acids play an important role in combating oxidative stress in the human body by maintaining a balance between oxidants and antioxidants (Temple, 2000). Thus, consumption of large amounts of buckwheat bran is considered to have significant nutritional or medicinal benefits. However, refined buckwheat flour is milled by removing most fiber and mineral and phenolic compounds in the bran and germ. Therefore a gradual reduction milling system has been developed to improve the nutritional constitution in flour. This method separates whole wheat grains into various fractions by weight from outer to inner parts (Maeda et al., 1999). The quality of these flour fractions was improved by the increase of nutritive components from the bran and germ which are rich in dietary fiber, resistant starch, vitamins, minerals and microconstituents. For the gradual milling of buckwheat, the grains were first broken and the husks were removed by a dehulling apparatus with disks forming groats. Then the groats were cut into 7–8 pieces and the cut germ is removed. The remaining groats with endosperm and bran were milled to various flour fractions by their weight. The objective of this chapter is to evaluate the nutritional composition, phenolic compounds and their antioxidant capacity in each fraction and the use of these fractions for noodle making.

1. Characterization of Polished-Grading Buckwheats

1.1. Chemical Composition of Polished-Grading Buckwheats

In previous study, the whole Mankan buckwheat grains were used to mill into sixteen flour fractions from inner to outer layers (Hung et al., 2007). Namely, FS, F1, F2 and F3 were the fractions graded from the innermost part to the outer part of grain. This means FS was the central part of grain and the other fractions approached the outer layer of the bran, respectively. Furthermore, FS, F1, F2 and F3 fractions were separated into various smaller fractions.

The fraction number 1 (FS-1) was milled from the most central part of the endosperm and the other fractions as indicated by the increase in the fraction numbers were the parts closer to the outer layer of the bran. The yields of FS, F1, F2 and F3 were 40.95%, 20.54%, 16.67% and 21.84%, respectively (Hung et al., 2007). Moisture contents of flours were lower in the outer fractions, whereas protein and ash contents increased from inner to outer part of grains (Table 4-1).

The whiteness (L*) of the flour fraction decreased from the inner to the outer parts, whereas the chromaticity coordinate (b*) of the outer part was higher than that of the inner part. These results showed that the inner fractions which contained mostly starch were much whiter than the outer fractions, whereas the higher phenolic compounds, protein and dietary fiber of the outer fractions contributed to the dark-gray color of flours.

Table 4-1. Characteristics and compositions of various graded buckwheat flour fractions (adapted from Hung *et al.*, 2007)

Fraction number	Fraction name	Yield (%)	Main constituents (%Wt)			Color of flour[a]		
			Moisture	Ash	Protein	L*	a*	b*
1	FS-1	17.56	14.0	0.188	3.15	92.34	-0.94	6.43
2	FS-2	3.19	13.9	0.160	3.28	92.56	-0.92	6.35
3	FS-3	9.92	14.1	0.176	3.60	92.56	-0.83	6.26
4	FS-4	9.13	13.4	0.247	3.96	91.90	-0.83	6.26
5	F1-1	11.41	14.0	0.224	3.90	90.79	-0.78	6.88
6	F1-2	9.13	13.5	0.298	4.34	92.17	-0.87	6.66
7	FS-5	1.15	13.7	0.483	5.93	91.31	-0.77	7.02
8	F2-1	9.18	13.0	0.845	8.03	87.75	-1.16	9.21
9	F2-2	1.05	12.2	1.192	10.42	86.28	-1.31	10.17
10	F2-3	1.71	12.9	1.825	15.24	85.96	-1.16	10.39
11	F2-4	1.31	11.0	2.178	17.46	79.37	0.40	9.05
12	F2-5	3.42	11.1	3.422	26.57	82.95	-0.46	10.46
13	F3-1	3.59	10.6	4.145	31.75	80.34	-0.63	11.87
14	F3-2	4.56	10.7	4.644	35.14	79.77	-0.77	12.22
15	F3-3	4.56	10.4	5.401	40.06	75.36	-0.80	13.35
16	F3-4	9.13	10.4	5.401	40.06	76.57	-0.79	12.88

[a] L, Lightness; a and b, Chromaticity coordinates (a* green to red and b* blue to yellow).

1.2. Phenolic Compounds and Antioxidant Capacity of Polished-Grading Buckwheats

Like wheat, phenolic compounds in buckwheat flours are also extracted in free and bound forms. The free form of phenolic compounds is easily extracted by organic solvents like ethanol, methanol or acetone, whereas the bound phenolic compounds are extracted by acetone-ethyl acetate solution after hydrolysis by alkaline solution. The total phenolic compounds in the free and bound extracts are determined using Folin-Ciocalteu's reagent method. The phenolic contents of both free and bound phenolic extracts of buckwheat flour fractions significantly increased in the order from the fraction number 1 (phenolics less rich fraction) to the fraction number 16 (phenolics rich fraction) (Figure 4-2) (Hung and Morita, 2008). Total phenolic contents of the F3-3 and F3-4 fractions, which were the phenolics rich fractions, were 30-fold higher than those of the FS-1 and FS-2 fractions, which were the phenolics less rich fractions.

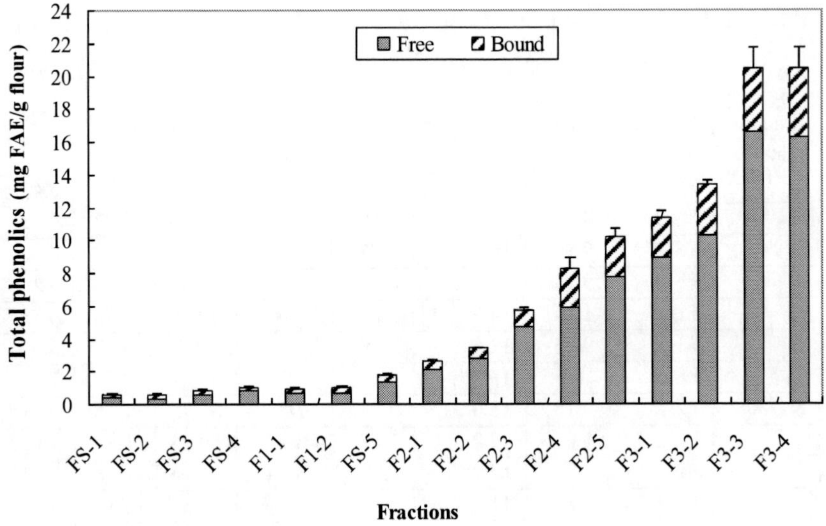

Figure 4-2. Free and bound phenolic contents of buckwheat graded milling flours. Values are means of triplicate measurements (adapted from Hung and Morita, 2008).

The results indicate that the bran and germ of buckwheat grains contain high amounts of phenolic compounds like other cereals such as wheat, barley, etc. The main phenolic acids, gallic, p-hydroxybenzoic, p-coumaric and ferulic acids, and the main flavonoids, such as rutin and quercetin, were also found in the free and bound phenolic extracts. The rutin in buckwheat grain was mostly in free form and was easily extracted by using ethanol or methanol compared with other flavonoids which existed in bound forms associated with cell wall materials (Hung and Morita, 2008).

DPPH is one of a few stable and commercially available organic nitrogen radicals and has a maximum UV-Vis absorbance at 515 nm. Measurement of antioxidant capacity is based upon the ability to reduce the DPPH radical (Huang et al., 2005; Prior et al., 2005).

The scavenging of the stable DPPH radical is widely used to evaluate antioxidant activity of phenolic compounds extracted from fruit and vegetable, cereal grain, wine, etc. The scavenging capacity of DPPH radical gradually increased in the order from the fraction number 1 (phenolics less rich fraction) to the fraction number 16 (phenolics rich fraction) for both free and bound phenolic extracts (Figure 4-3). After 30 min, the DPPH scavenging of FS-1 (phenolics less rich fraction) were 5.5 and 1.1%, whereas the scavenging of F3-4 (phenolics rich fraction) were 94.4 and 34.2%, for the free and bound phenolic extracts respectively.

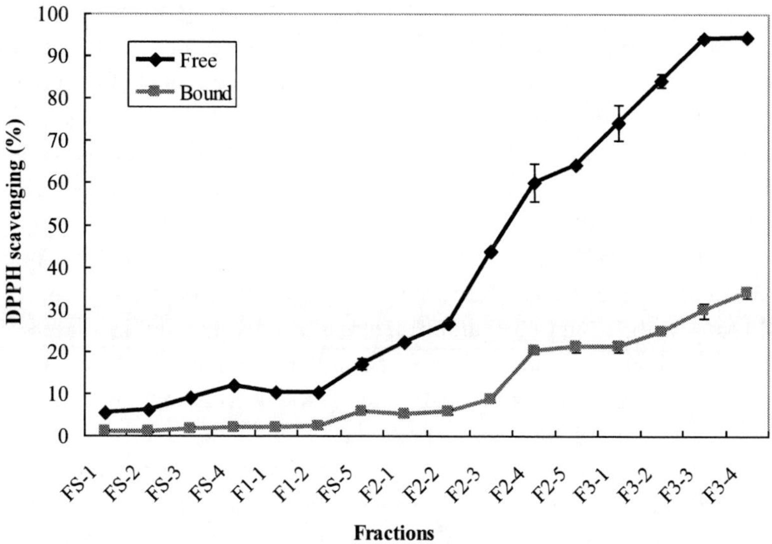

Figure 4-3. DPPH (1,1-diphenyl-2-picrylhydrazyl) radical scavenging capacity of free and bound phenolic extracts from buckwheat graded milling flours. Concentration of DPPH = 0.15 mM. Values are means of triplicate measurements (adapted from Hung and Morita, 2008).

In the DPPH assay, the bound phenolic extracts exhibited significantly lower scavenging than the free phenolic extracts, which is contrary to the other cereal grain extracts (Adom and Liu, 2002; Liyana-Pathirana and Shahidi, 2007). These results indicate that the antioxidant capacity of the phenolic extracts was caused by higher total phenolic and rutin contents in the free phenolic extracts than the bound phenolic extracts of buckwheat flour fractions. As a result, the outer layers of buckwheat grains with higher amount of phenolic compounds along with higher amounts of protein, lipid, ash and dietary fiber are considered to be good materials for cereal-based food processing with significant health benefits.

2. Application of Polished-Grading Buckwheat for Noodle Making

Eight buckwheat flour fractions (fractions FS-1, FS-3, F1-1, F1-2, F2-1, F2-3, F2-5, F3-1) were used to substitute for 40% of wheat flour to make soba noodles. The noodles were made from 250 g of the flours, 2% of sodium chloride (g g^{-1} flour) and 34% of water (g g^{-1} flour) using a pasta machine. An exception was the fraction F3-1 which could not be extruded until the amount of water decreased by 5%. Optimum cooking time and color of uncooked and cooked soba noodles are given in Table 4-2. The optimum cooking time of soba noodles did not change for the substitution with FS fractions (FS-1 and FS-3), but the substitution with F1, F2 and F3 fractions decreased the optimum cooking time of noodles.

The substitution with FS and F1 fractions increased the L* value, but decreased b* value of both uncooked and cooked noodles. In contrast, the substitution with F2 and F3 fractions decreased L* value, but increased a* and b* values of both uncooked and cooked noodles as compared to the control. These results show that the noodles substituted with the inner buckwheat fractions were whiter but less yellow than that of the control, whereas the noodles substituted with the outer buckwheat fractions were darker and more yellow than that of the control because they contained a lot of phenolic compounds. Rheological properties of cooked noodles substituted with the buckwheat fractions are shown in Figs. 4-4 and 4-5. The FS and F1 fraction substitutions for wheat flour 1CW increased firmness of cooked noodles. The substitution with F2-1 fraction also increased the firmness of noodle as compared to the control, whereas the substitution with F2-3 and F2-5 had similar firmness of noodle to the control.

Table 4-2. Cooking time and color of noodle made from various graded buckwheat flour fractions (adapted from Hung et al.**, 2007)**

Fraction number	Fraction name	Optimum cooking time (min)	Color of noodle[a,b]					
			Uncooked			Cooked		
			L*	a*	b*	L*	a*	b*
Control	1CW	4.5	67.5	+ 0.9	+ 19.6	64.4	- 2.2	+ 9.0
1	FS-1	4.5	75.7	+ 0.5	+ 15.0	66.3	- 1.8	+ 5.4
3	FS-3	4.5	74.2	+ 1.4	+ 15.8	66.4	- 1.5	+ 7.1
5	F1-1	4.0	72.7	+ 0.9	+ 15.3	65.7	- 1.2	+ 6.9
6	F1-2	4.0	73.8	+ 1.1	+ 17.2	66.4	- 1.6	+ 6.3
8	F2-1	4.0	65.6	+ 1.4	+ 18.3	66.3	- 0.8	+ 8.6
10	F2-3	3.5	58.7	+ 3.5	+ 20.2	63.0	+ 0.8	+ 9.0
12	F2-5	3.0	48.5	+ 5.9	+ 18.9	57.5	+ 3.3	+ 9.0
13	F3-1	3.5	56.3	+ 5.6	+ 19.9	56.5	+ 2.8	+ 7.0

[a] Color of noodles were measured before and after boiling at optimum cooking time.

[b] L, Lightness; a and b, Chromaticity coordinates (a* green to red and b* blue to yellow).

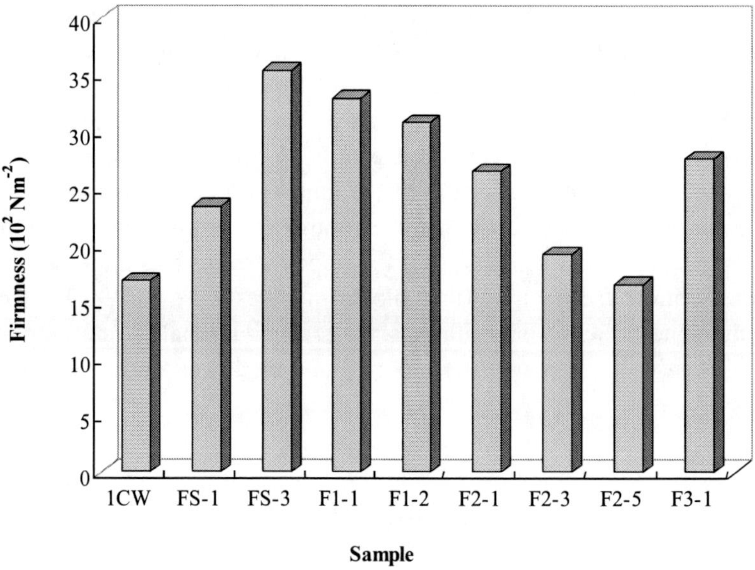

Figure 4-4. Firmness of cooked noodles made from the buckwheat flour fraction substitution (adapted from Hung et al., 2007).

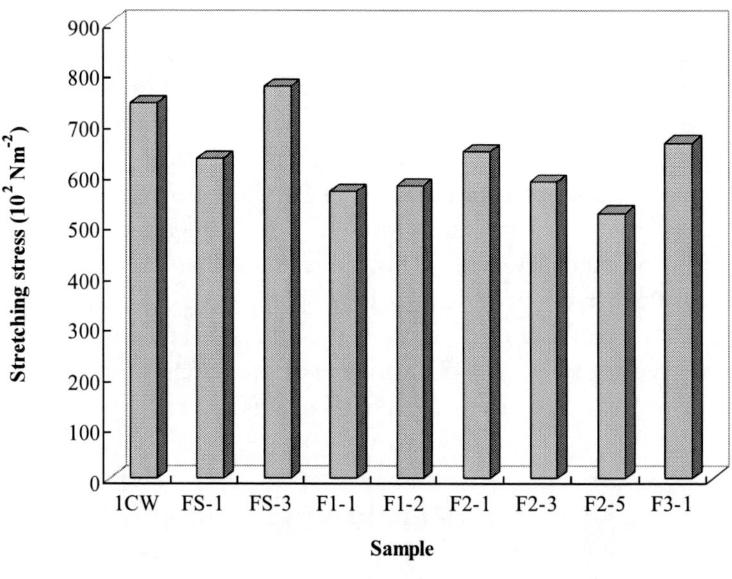

Figure 4-5. Stretching stress of cooked noodles made from the buckwheat flour fraction substitution (adapted from Hung et al., 2007).

For the substitution with fraction F3-1 (water absorption reduced by 5% to be able to extrude) the firmness of cooked noodle was higher than that of the noodles substituted with F2-3 and F2-5 because of the reduction of water. The stretching stress of cooked noodles with all buckwheat fraction substitutions was lower than that of the control, except FS-1 fraction which was similar to the control. Thus, the substitution of the inner buckwheat fractions made the noodles harder and less elastic than the control. In contrast, the substitution of the outer buckwheat fractions made the noodles softer and more elastic than that of the inner fractions but the color of noodles were darker. The noodles with the inner fraction substitutions have an additional benefit for human health because it was previously reported to have lower amount of allergenic protein.

CONCLUSION

The gradual milling method was developed to increase the nutritive values and the noodle making quality of buckwheat flour by using whole buckwheat grains for milling. The protein and ash contents of buckwheat flour fractions increased in the order from the inner to the outer fractions. The inner fractions which contained mostly starch were much whiter than the outer fractions, which contained higher amounts of phenolic compounds, protein and dietary fiber and exhibited stronger antioxidant capacity. The different composition of the fractions affects the quality of cooked noodles made from these buckwheat flour fractions. The optimum cooking time and whiteness of uncooked and cooked noodles decreased from the inner fraction to outer fraction, whereas the noodles made from the inner fractions were harder and less elastic than those from the outer fractions. Thus, the composition and characteristics of buckwheat flour fractions varied from the outer parts to the inner parts. These flours could be used for food processing to increase the nutritive values such as protein, dietary fiber, phenolic compounds in the outer fractions or reduce the allergenic proteins by using the inner fractions.

REFERENCES

Adom, K.K., and Liu, R.H. (2002). Antioxidant activity of grains. *Journal of Agriculture and Food Chemistry*, 50, 6182-6187.

Craig, S.A.S., Holden, J.F., Troup, J.P., Auerbach, M.H., and Frier, H.I. (1998). Polydextrose as soluble fiber: Physiological and analytical aspects. *Cereal Food World*, 43, 370-376.

Franschi, M.L.P. de, Salgado, J.M., and Leitao, R.F.F. (1994). Chemical, nutritional and technological characteristics of buckwheat and non-prolamine buckwheat flour in comparison of wheat flour. *Plant Foods for Human Nutrition*, 46, 323-329.

Holasova, M., Fiedlerova, V., Smrcinova, H., Orsak, M., Lachman, J., and Vavreinova, S. (2002). Buckwheat - the source of antioxidant activity in functional foods. *Food Research International*, 35, 207–211.

Huang, D., Ou B., and Prior, R.L. (2005). The chemistry behind antioxidant capacity assays. *Journal of Agriculture and Food Chemistry*, 53, 1841-1856.

Hung, P.V., Maeda, T., Tsumori, R., and Morita, N. (2007). Characteristics of fractionated flours from whole buckwheat grains using gradual milling system and their application for noodle making. *Journal of the S cience of Food and Agriculture*, 87, 2823-2829.

Hung P.V., and Morita N. (2008). Distribution of phenolic compounds in the graded flours milled from whole buckwheat grains and their antioxidant capacities. *Food Chemistry* 109, 325-331.

Joshi, B.D., Rana, R.S. (1995). Buckwheat (*Fagopyrum esculentum*). In: Williams JT (Ed) *Cereal and Pseudocereals*. Chapman and Hall, Lodon, pp 85-127.

Liyana-Pathirana, C.M., and Shahidi, F. (2007). The antioxidant potential of milling fractions from breadwheat and durum. *Journal of Cereal Science*, 45, 238-247.

Maeda, T., Ohkura, M., and Morita, N. (1999). Characterization of wheat flour graded by polishing and its application to breadmaking. *Journal of Applied Glycoscience*, 46, 413–422.

Marshall, H.G., and Pomeranz, Y. (1982). Buckwheat: Description, breeding, production, and utilization. In: Pomeranz Y (Ed) *Advances in Cereal Science and Technology* (Vol. 5). AACC International, St. Paul, MN, pp 157-210.

Mazza, G., and Oomah, B.D. (2005). Buckwheat as a food and feed. In: Abdel-Aal E, Wood P (Eds) *Specialty Grains for Food and Feed*. AACC International, St. Paul, MN, pp 375-393.

Ohnishi, O. (1998). Search for the wild ancestor of buckwheat. III. The wild ancestor of cultivated common buckwheat, and of tartary buckwheat. *Economic Botany*, 52, 123-133.

Pomeranz, Y. (1983). Buckwheat: structure, composition, and utilization. *Critical Reviews in Food Science and Nutrition*,19, 213–258.

Prior, R.L., Wu, X., and Schaich, K. (2005). Standardized methods for the determination of antioxidant capacity and phenolics in foods and dietary supplements. *Journal of Agricultural and Food Chemistry*, 53, 4290-4302.

Temple, N.J. (2000). Antioxidants and disease: more questions than answers. *Nutrition Research*, 20, 449-459.

Topping, D.L, and Clifton, P.M. (2001). Short-chain fatty acids and human colonic function: roles of resistant starch and nonstarch polysaccharides. *Physiological Reviews*, 81, 1031-1064.

Yoshimoto, Y., Egashira, T., Hanashiro, I., Ohinata, H., Takase, Y., and Takeda, Y. (2004). Molecular structure and some physicochemical properties of buckwheat starches. *Cereal Chemistry*, 81, 515-520.

Udesky, J. (1988). *The book of soba,* Kodansha International, New York.

Watanabe, M., Ohshita, Y., and Tsushida, T. (1997). Antioxidant compounds from buckwheat (*Fagopyrum esculentum* Moench) hulls. *Journal of Agricultural and Food Chemistry*, 45, 1039-1044.

Wijngaard, H.H., and Arendt, E.K. (2006). Buckwheat. *Cereal Chemistry*, 83, 391-401.

In: Advances in Cereal and Pseudocereal ... ISBN: 978-1-62618-347-6
Editors: N. Morita, P.V. Hung et al.

Chapter 5

GERMINATED BUCKWHEAT FOR FUNCTIONAL FOODS

Naofumi Morita[1], Kazuyoshi Miyake[2], Tomoko Maeda[3] and Pham Van Hung[4]

[1] Department of Food Packaging Technology, Toyo College of Food Technology, Mimami-Hanayashiki, Kawanishi, Hyogo, Japan
[2]Miyake Flour Milling Co., Ltd., Momodani, Ikuno-ku, Osaka, Japan
[3]Department of Science, Technology, and Human Life, Graduate School of Education, Hyogo University of Teacher Education, Shimokume, Kato, Hyogo, Japan
[4]School of Biotechnology, International University, National University in HoChiMinh City, HoChiMinh City, Vietnam

ABSTRACT

Buckwheat grains were gradually milled from the inner to the outer layers. The characteristics of 5 types of buckwheat flour obtained from milling [(special flour (FS), 1st-3rd grade flours (F1-F3) and seed coat flour (FBB)] were determined. The protein and ash content of all buckwheat fractions increased as milling progressed from the inner fraction to the outer fraction. The F3 fraction, which includes bran and the germ, was the richest in nutritional value. During germination at 25°C, both gamma-amino butyric acid (GABA) levels and the enzymatic activity of α-amylase and protease showed a greater increase than during wheat germination. To decrease the allergenicity of buckwheat grains, buckwheat germination was performed and the characteristics of the grains were assessed. Germinated buckwheat grains were fermented to generate the processed foods, such as soba *natto* and soba *miso* paste. For soba *miso* paste, the paste became dark with red and yellow tints due to the Maillard browning reaction, the pH value decreased gradually (pH 6.25 to 5.6) after storage for 60 days, and the total titratable acidity (TTA) value increased (3.8 to 9.5) according to the length of fermentation. During fermentation, the amounts of GABA produced in soba *natto* and *miso* paste 48 hr and 60 days of fermentation, respectively, were 3.3 and 1.7 times higher than at the start of fermentation. SDS-PAGE analysis of albumin and globulin from soba *natto* during fermentation indicated that high molecular weight proteins (45-97 kDa) were less abundant after 48 hr of fermentation, and low molecular weight (LMW) proteins were generated. Allergenic proteins were identified with IgE immunoblotting, and the majority of the allergenic 12 and 22 kDa proteins from albumin and globulin were degraded to LMW proteins. The HMW albumin and globulin proteins from soba *miso* paste decomposed after 60 days of fermentation, and the ability of the proteins to bind to IgE also decreased. For commercial-scale germination of buckwheat grains, heating and humidifying of moist air methods was applied rather than soaking directly in water.

As a result, germinating buckwheat might improve the properties of the grain, leading to a higher nutritional value, improved flavor, and decreased allergenicity.

INTRODUCTION

The word "buckwheat" contains "wheat," but buckwheat is not a cereal. The plant is classified as a pseudocereal and a dicotyledonous plant. The production of buckwheat grain has many drawbacks compared to rice or wheat due to gramineous plants. Therefore, cultivation today is limited to highland or alpine valleys where rice and wheat cannot be grown. However, among grain crops, buckwheat is gaining popularity because it contains bioactive components with beneficial health properties. Additionally, buckwheat proteins are composed of well-balanced amino acids with a high amino acid score and have anti-cholesterolemic properties (Tomotake et al., 2000). Buckwheat starches are less easily digestible, affording a low glycemic index (Skrabanja et al., 2001). Furthermore, many of the health benefits of the buckwheat grain are correlated to the presence of gamma-amino butyric acid (GABA), the flavonoids rutin and quercetin, and other phenolics. Rutin has been reported to alleviate many modern health disorders caused by the nutritional habits currently prevailing in western countries. The beneficial properties of rutin include antimutagenic activity (Undeger, 2004), protective effects against the development of diabetes (Srinivasan et al., 2005), antioxidative properties (Oomah & Mazza, 1996, Hung & Morita, 2008), a well-balanced amino acid composition with a high biological value (Pomeranz, 1983), and a decrease in plasma cholesterol levels (Kayashita et al., 1997). However, another property of buckwheat critical to human health is the presence of proteins that cause a hypersensitive allergic reaction.

Since the development of the modern milling apparatus, refined buckwheat flour contains little of the fiber, minerals and phenolic compounds found in the bran and germ. Therefore, we are obliged to eat buckwheat flour that is less nutritious than it could be. Recently, a gradual reduction milling system was developed to improve nutritional content and to remove allergenic proteins from flour. We also germinated buckwheat grains to decrease the allergenic protein content and to increase the nutritional properties of buckwheat. The purpose of the present study was to determine the characteristics of various graded fractions obtained by milling the germinated buckwheat cultivar Mancan and also to characterize changes in the properties of germinated buckwheat flour. These results can be applied to develop milling procedures for buckwheat grains that yield buckwheat products with low- or no-allergenicity.

Grain and Flour

The buckwheat grain used for germination was the cultivar Mancan, which was imported from China and supplied by Miyake Flour Milling Co., Ltd. (Osaka, Japan). The buckwheat grain was gradually milled from the innermost portion of the grains; the five fractions FS, F1, F2, F3 and FBB were obtained in order from the innermost to the outermost fractions, as shown in Fig. 5-1.

Flour quality was tested using ICP (Inductively Coupled Plasma) analysis to assess inorganic compounds and amino acid content. Scanning electron microscopy (SEM) was used to observe the starch granules of the different fractions of flour prepared from germinated grains. Whole buckwheat grains (groats) were soaked in water for 1 hr at 25°C, then germinated for the specified time with a relative humidity of 85% at 25°C. For comparison, the wheat grain ICW was germinated under the same conditions, except the temperature was 30°C. After germination for the specified time interval, the buckwheat grains were collected, frozen, and lyophilized.

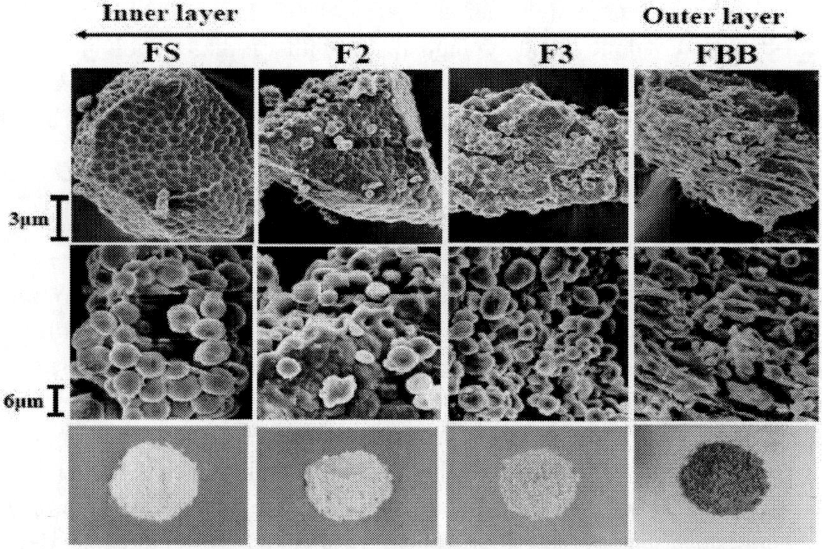

Figure 5-1. SEM images of graded buckwheat flours (adapted from Miyake et al., 2004). FS (×1000, 2500); F2 (×1000, 2500); F3 (×1000, 2500); FBB (×1000, 2500).

Characteristics of Germinated Buckwheat Flour

Table 5-1 shows a proximate analysis of the buckwheat flour. The moisture content of buckwheat flours decreased in the order FS, F1, F2, F3 and FBB, whereas the amount of ash increased in the same order, except for the FBB fraction. The amount of protein, lipid and ash increased from the inner fraction FS to the outer fraction F3. FBB, the outermost fraction, contained the largest amount of carbohydrates.

The mineral content of the inner fractions FS, F1 and F2 contained smaller amounts of Ca, whereas the outer F3 and FBB fractions contained larger amounts (Table 5-2). All fractions contained large amounts of K, Mg and P, and the FBB fraction in particular contained the largest amount of S among all the fractions tested.

The amount of phytic acid (IP6) in various germinated flour samples was measured using the spectrophotometric method at 500nm (Latta and Eskin, 1980). During germination for 24 hr, the amount of IP6 detected from buckwheat did not change appreciably (0.44-0.47%), whereas the amount measured in wheat increased significantly (0.07-0.35%), reaching approximately 5 times the starting amount (data not shown).

Table 5-1. Proximate analyses of buckwheat fractions (adapted from Miyake et al., **2004)**

Sample	Moisture	Ash	Protein	Lipid	Carbohydrate
			(%)*		
FS	14.4	0.2	3.8	2.6	79.0
F1	14.3	0.5	5.5	3.0	76.7
F2	12.4	3.1	23.7	5.9	55.0
F3	10.9	5.2	38.1	8.8	37.0
FBB	7.2	2.6	4.0	2.8	83.4

FS, F1, F2, F3 and FBB were gradually milled from the innermost to the outermost fraction of a whole buckwheat grain. FS, special flour; F2-F3, 2nd~3rd grade flours; FBB, seed coat flour.

* Values were calculated on a dry flour basis.

Table 5-2. Mineral components of graded buckwheat flours (adapted from Miyake et al., **2004) (mg/100g of dry flour)**

Element	FS	F1	F2	F3	FBB
B	0.27	0.18	0.79	1.43	0.54
Ca	0.001	0.001	0.002	43.4	117.6
Cu	0.13	0.14	0.32	0.73	0.33
Fe	0.30	0.48	2.73	5.03	1.27
K	140.7	190.4	1085.5	2208.9	905.6
Mg	22.8	34.8	228.3	432.2	85.4
P	52.6	92.3	597.0	1129.0	66.7
Pd	0.04	0.04	0.04	0.04	0.05
Rb	0.95	1.08	3.97	7.60	1.68
S	1.08	0.89	0.35	0.59	2.40
Se	0.06	0.11	0.10	0.01	0.08
Sn	0.00	0.01	0.00	0.02	0.00
Zn	0.27	0.42	2.34	4.41	0.52

Abbreviations are the same as in Table 5-1.

Amino Acid Composition of Graded Flours and Germinated Grains

From the amino acid composition of the various buckwheat fractions shown in Table 5-3, it is apparent that the F3 fraction containing the germ and seed coat had larger quantities of amino acids compared to the other fractions. In addition, F3 clearly contained a larger amount of GABA than the other samples. Although the FBB fraction is similar to the F3 fraction in relative amino acid composition, the FBB fraction contained strikingly decreased quantities of amino acids compared to F3. However, the amount of GABA in FBB was higher than the FS, F1 and F2 fractions.

Table 5-3. Amino acid components in graded buckwheat flours* (adapted from Miyake et al., **2004) (mg/100g of dry flour)**

Amino acid	FS	F1	F2	F3	FBB
Asp	61	467	1603	2614	154
Thr	239	218	574	1052	150
Ser	302	312	991	1624	215
Asn	13	21	581	0	0
Glu	1535	1246	4964	8471	1461
Pro	195	99	1029	2133	189
Gly	646	567	1179	2027	333
Ala	505	300	900	1349	155
Val	301	33	887	1412	102
Cys	61	73	604	1016	52
Met	25	168	445	668	70
I-leu	250	162	655	1199	1.9
Leu	454	374	1275	2181	196
Tyr	188	177	473	873	13
Phe	311	215	747	1211	90
GABA	9	3	87	310	89
Orn	128	20	66	15	1
Lys	500	370	1039	1800	224
His	184	767	362	547	28
Arg	641	419	2119	3870	91
Total amounts	7107	6531	21774	34954	4354

Abbreviations are the same as in Table 5-1.

*All data were collected by hydrolysis with hydrochloric acid.

Germination of buckwheat grain generated higher levels of amino acids, especially GABA, which increased 2.5 times after incubation for 24 hr compared to without germination, as shown in Table 5-4. By comparison, GABA increased 1.5 times under the same conditions in germinating wheat grains.

Table 5-4. Changes to free amino acid content of buckwheat and wheat grains during germination (adapted from Miyake et al.**, 2004)**

Amino acid	Buckwheat (mg/100g of dry flour)				Wheat (mg/100g of dry flour)			
	Control	G-8	G-16	G-24	Control	G-8	G-16	G-24
THR	13.4	18.2	21.3	29.5	4.1	5.6	12.8	12.6
VAL	5.5	4.7	6.5	10.4	5.7	6.8	15.2	16.1
MET	8.8	9.9	10.3	13.2	2.1	5.9	5.1	5.3
LEU	11.7	13.2	13.0	16.9	3.8	7.2	16.8	15.9
TYR	8.7	11.2	9.9	10.7	4.0	3.6	11.5	12.5
PHE	10.2	11.9	13.2	16.8	3.3	4.8	11.6	12.0
HIS	5.6	10.6	13.7	23.2	2.7	4.0	9.2	8.5
LYS	9.6	10.4	13.2	19.6	9.2	10.4	12.8	17.6
GLU	72.4	75.4	105.6	133.4	36.7	59.6	94.1	82.6
GLN	16.5	27.3	44.6	88.4	13.7	5.0	34.5	34.6
GABA	12.4	11.6	22.5	28.7	5.2	3.6	6.9	7.9
Total	174.8	204.4	273.8	390.8	90.5	116.5	230.5	225.6

In terms of the α-amylase and protease activity during germination, buckwheat is more active than wheat grain. α-Amylase from buckwheat showed a peak in activity after incubation for 16 hr, and protease activity increased steadily for the length of incubation, as reported by Kikunaga and Takahashi (1992).

As described in the results above, each graded flour fraction possessed different flour qualities and different functional and nutritional properties. It is well known that the amino acid score of buckwheat is much higher than rice or wheat because buckwheat contains a large amount of lysine and tryptophan. Therefore, germination experiments were performed with the expectation that the level of allergenic substances in buckwheat would decrease during the germination process. In fact, *Rhizopus oligosporus* has been known to increase the functional properties of tempeh, which is made from buckwheat or soybean, and was very effective at reducing the allergenic protein composition of buckwheat, as reported by Handoyo et al. (2006) and Kobayashi et al. (2004). For practical food applications, buckwheat grains were used for the preparation of soba *natto* and *miso* paste, both typical and traditional foods in Japan. *Natto* is highly regarded worldwide as a healthy food.

Preparation of Soba *Natto* and *Miso* Paste Using Germinated Buckwheat

Germinated and autoclaved buckwheat was used for the preparation of soba *natto* and *miso* paste. Soba *natto* was prepared by inoculating the buckwheat with *Bacillus natto* spores. Fermentation occurred for 0, 12, 24, 36 or 48 hr at 40°C using a commercial "Domestic *natto* maker" apparatus according to the manufacturer's instructions (Tokyo Unicom Co. Ltd., Tokyo, Japan). Soba *miso* paste was prepared using pre-germinated buckwheat (450g), malted rice (300g), salt (110 g) and water (50g), then stored in an incubator at 30°C for 0, 15, 30, 45 or 60 days.

Characteristics of Soba *Natto* and *Miso* Paste Made from Germinated Buckwheat

During soba *natto* fermentation, the protein content noticeably decreased; therefore, the characteristics of the proteins in soba *natto*, including molecular weight and stability, were changed during fermentation.

For soba *miso* paste, the appearance of the paste was considerably changed to a darker color due to the Maillard browning reaction. After storage for 60 days, the soba *miso* paste was dark with a red and yellow tint. The pH value decreased gradually (pH 6.25 to 5.6), and the total titratable acidity (TTA) value increased (3.8 to 9.5) according to the length of fermentation.

Amino Acid Changes in Soba *Natto* and *Miso* Paste

During fermentation, the amount of GABA produced in soba *natto* and *miso* paste clearly increased, particularly the amounts of GABA in soba *natto* and *miso* paste after 48 hr and 60 days of fermentation, which were 3.3 and 1.7

times higher than at the start of fermentation, respectively (Tables 5-5 and 5-6).

Table 5-5. Amino acid composition of soba *natto* during fermentation (adapted from Miyake et al.**, 2006) (mg/100mL)**

Amino acid	Fermentation time (hr)				
	0	12	24	36	48
Asp	3.4	1.3	4.1	7.4	6.7
Thr	7.2	1.9	2.2	1.1	3.3
Ser	7.8	4.6	2.5	1.0	1.8
Asn	3.3	1.3	2.6	2.2	0.8
Glu	47.8	17.4	12.3	7.6	12.8
Pro	4.7	1.8	5.7	3.6	7.8
Gly	3.8	2.1	0.7	0.1	0.6
Ala	9.9	10.1	4.5	1.6	3.9
Val	4.3	2.9	3.4	2.8	3.6
Cys	8.7	6.0	12.0	6.7	12.2
Met	1.0	0.7	1.3	1.4	5.1
I-leu	42.0	34.2	32.8	17.4	19.9
Leu	16.4	11.5	22.8	3.4	9.9
Tyr	3.5	3.0	4.0	4.4	5.4
Phe	3.8	2.9	4.0	4.9	5.4
GABA	2.6	7.6	6.9	6.5	8.6
Etha	3.6	4.1	12.1	5.8	18.3
Ammo	5.9	4.0	11.4	9.3	15.7
Orn	0.3	1.8	8.7	3.3	10.1
Lys	5.8	3.5	5.6	5.3	5.5
His	12.8	9.5	11.5	10.2	13.8
Arg	31.6	17.2	10.4	3.6	3.5
Total	230.5	149.5	181.6	109.5	174.8

In addition, glutamic acid levels in soba *natto* decreased (0.27 times), although in soba *miso* paste these levels increased (5.7 times). Because GABA is generated from glutamic acid by glutamate decarboxylase, the decrease in glutamic acid levels in soba *natto* during fermentation might be caused by GABA formation. In the fermentation of soba *miso* paste, the amount of glutamic acid produced was higher than the amount consumed by GABA formation during fermentation. The total amount of amino acids in soba *miso* paste also increased, but in soba *natto* the amino acid levels decreased during fermentation. Therefore, in the production of soba *miso* paste, new amino acids were produced rather than consumed during fermentation; however, the amino acids were substantially consumed during soba *natto* fermentation, resulting in lower levels when compared to the start of fermentation. This result correlated with decreasing protein levels during fermentation.

Table 5-6. Amino acid composition of soba *miso* paste during fermentation (adapted from Miyake et al., 2006) (mg/100mL)

Amino acid	Fermentation time (days)				
	0	15	30	45	60
Asp	0.6	7.0	22.3	31.0	21.9
Thr	1.2	7.9	10.0	10.8	7.8
Ser	2.1	9.1	19.8	24.1	18.4
Asn	1.0	9.1	10.1	9.6	12.0
Glu	11.7	53.5	72.4	81.8	66.8
Pro	3.3	8.3	10.3	11.1	8.7
Gly	0.3	7.4	17.0	21.6	15.9
Ala	14.6	26.8	49.3	52.1	43.6
Val	1.4	8.4	18.8	21.0	15.8
Cys	5.5	9.6	7.6	9.4	19.4
Met	0.7	3.1	1.9	2.0	3.9
I-leu	18.7	50.8	19.7	20.6	72.7
Leu	10.2	34.1	30.1	31.4	51.8
Tyr	2.3	7.5	14.9	16.1	11.3
Phe	0.5	8.4	9.5	7.2	8.7
GABA	6.5	7.9	4.9	4.2	10.9
Etha	34.9	41.3	17.2	23.9	69.1
Ammo	7.0	11.0	8.2	8.9	16.6
Orn	12.8	5.5	7.6	7.0	5.0
Lys	13.5	44.9	46.3	49.0	58.2
His	1.8	8.2	12.3	13.3	11.2
Arg	11.1	53.9	68.1	73.9	73.4
Total	161.8	423.8	478.3	530.1	623.1

DEGRADATION OF ALLERGENIC PROTEINS IN BUCKWHEAT DURING FERMENTATION

SDS-PAGE analysis of albumin and globulin proteins from soba *natto* during the fermentation time course is shown in Fig. 5-2.

The protein composition at the start of fermentation (0 hr of fermentation) shows proteins of various molecular weights, but high molecular weight (HMW) proteins were degraded during the time course of fermentation. The protein bands corresponding to albumin (45-97 kDa) at 48 hr decreased compared to 0 hr. For globulin, a protein band of approximately 45 kDa was faint, but low molecular weight (LMW) proteins of approximately 22 kDa were produced after fermentation for 48 hr, compared to the sample at the start of fermentation (0 hr). These results suggest that *B. natto* rapidly degraded the HMW proteins and utilized them for the production of amino acids or LMW peptides for microbial growth.

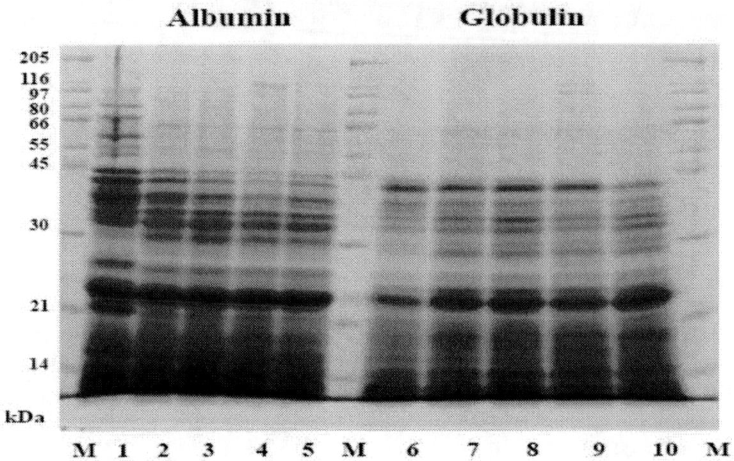

Figure 5-2. SDS-PAGE of albumin and globulin proteins extracted from soba *natto* samples after fermentation (adapted from Miyake et al., 2006). M is the molecular weight marker. 1, albumin (0); 2, albumin (12); 3, albumin (24); 4, albumin (36); 5, albumin (48); 6, globulin (0); 7, globulin (12); 8, globulin (24); 9, globulin (36); 10, globulin (48). Parenthetical values correspond to the fermentation time (hr).

The quantity of allergenic proteins in buckwheat samples was examined with immunoblotting analysis using an IgE that specifically binds to the proteins related to buckwheat allergy, as shown in Fig. 5-3. IgE bound most strongly to the 12 and 22 kDa albumin and globulin proteins in the control soba *natto* without fermentation (0 hr), and binding decreased after fermentation for 36 hr. According to a report by Yoshioka et al. (2004a, 2004b), the 22 kDa protein is a major allergen in Mancan buckwheat. Western blot analysis showed that the allergenic proteins were present in the control without fermentation (lanes 1 and 4 in Fig. 5-3) and were clearly albumin protein. However, these bands disappeared over the time course of fermentation (lanes 3 and 6 in Fig. 5-3). Therefore, in soba *natto* after 36 hr of fermentation, the allergenic albumin proteins were mostly degraded, and the allergenic globulin proteins were completely degraded to LMW proteins (amino acids or small peptides) (lanes 3 and 6 in Fig. 5-3).

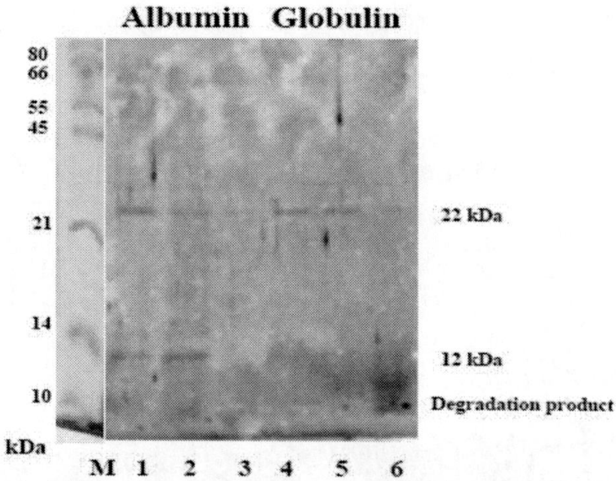

Figure 5-3. IgE immunoblotting of albumin and globulin proteins extracted from soba *natto* samples after fermentation (adapted from Miyake et al., 2006). M is the molecular weight marker. 1, albumin (0); 2, albumin (24); 3, albumin (36); 4, globulin (0); 5, globulin (24); 6, globulin (36). Parenthetical values correspond to the fermentation time (hr).

SDS-PAGE analysis of albumin, globulin, glutelin and prolamin in soba *miso* paste after fermentation for 0, 30 and 60 days are shown in Fig. 5-4.

The HMW protein bands from albumin, globulin and glutelin were weak and disappeared after 60 days of fermentation, although prolamin was an exception. Immunoblotting analysis of albumin and globulin after fermentation for 60 days demonstrated that the proteins of approximately 15 and 22 kDa disappeared after fermentation for 60 days (lanes 3 and 6 in Fig. 5-5).

From these results, it appears that the HMW albumin and globulin proteins from soba *natto* and *miso* paste were decomposed, and consequently, LMW proteins were produced during fermentation. In addition, the binding of IgE to these proteins, assessed by immunoblotting, became weaker over the long fermentation period, although the total amount of LMW proteins increased. Therefore, germinating buckwheat was effective, suitable and useful for the production of soba *natto* and *miso* pastes with low- or no-allergenicity. Furthermore, the germinated buckwheat can be used for the production of germinated-buckwheat products, such as soba *natto* and *miso* paste, which are a new type of fermented food distinct from ordinary *natto* and *miso* paste, which are made from soybeans.

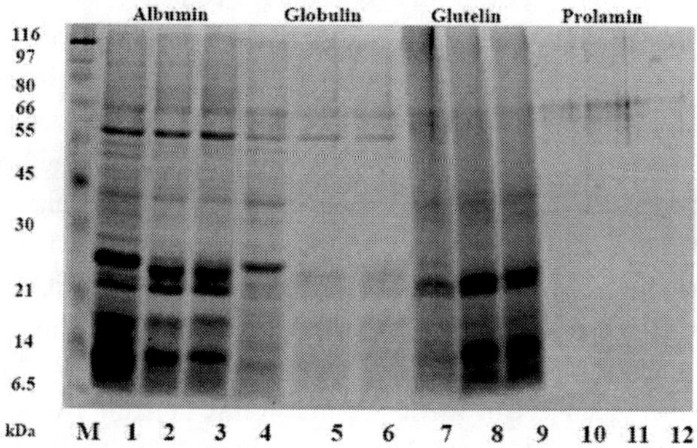

Figure 5-4. SDS-PAGE of proteins extracted from soba *miso* paste after fermentation (adapted from Miyake et al., 2006). M is the molecular weight marker. 1, albumin (0); 2, albumin (30); 3, albumin (60); 4, globulin (0); 5, globulin (30); 6, globulin (60); 7, glutelin (0); 8, glutelin (30); 9, glutelin (60); 10, prolamin (0); 11, prolamin (30); 12, prolamin (60). Parenthetical values correspond to the fermentation time (days).

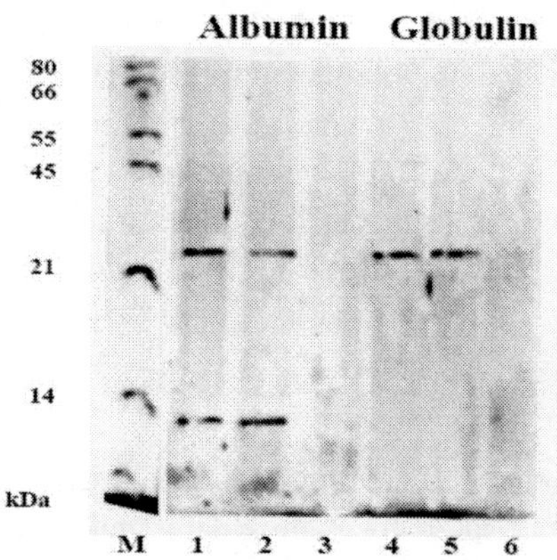

Figure 5-5. Immunoblotting analysis of albumin and globulin proteins extracted from soba *miso* paste after fermentation (adapted from Miyake et al., 2006). M is the molecular weight marker. 1, albumin (0); 2, albumin (15); 3, albumin (60); 4, globulin (0); 5, globulin (15); 6, globulin (60). Parenthetical values correspond to the fermentation time (days).

Improved Germination Methods for Cereal Grains

Recently, the preparation of germinated cereal grains avoided direct soaking of the grains in water because beneficial low molecular weight materials formed during germination were released into the solution. Therefore, minute amounts of water were used for germination. This method for GABA production was proposed as a fine moistening method by Satake Co. Ltd as the "Satake Co. method." (Satake et al., 2003; Satake, 2007). These methods were originally used for the germination of brown rice. The fine moistening method for brown rice hydrates the grains at 30°C and 19%-23% moisture (increased speed: 0.5%/hr for a grain moisture range below 17%, 0.5% to 1.2%/hr for a grain moisture over 17%, tempered for 12 hr) for 1 to 2 days. Using the "Satake Co. method," the grains contain 18mg GABA/100g dry basis, whereas after the water immersion method for 24 hr at 20~25°C, the grains contain 10~15mg GABA.

This method was further improved using a higher temperature (60-75°C), a shorter time (approximately 5 hr) with moist air (95% relative humidity) and termed the "heating and humidifying of moist air method" (Naka et al., 2011). The amount of GABA formed was approximately 3-fold (39.3mg) higher than control rie-Tartary buckwheat (13mg/100g dry basis).

Both the fine moistening method and the heating and humidifying of moist air method are convenient for the downstream milling process because extensive drying not required after germination. Additionally, these methods produce few heavy grain cracks, which is relevant to rice grain polishing.

Conclusion

Buckwheat has been used worldwide in processed foods, such as noodles, crepes, pasta, and some cakes for many years, and these products are usually made by mixing buckwheat flour with water and salt, followed by baking. In this study, we show that whole buckwheat grains can be used for processed foods without removing the germ, bran and seed coat (husk) after germination of the whole grain. Therefore, it is theoretically possible to utilize 100% of the grain, abrogating the need to discard byproducts such as the germ, bran and husk generated from the conventional system of buckwheat milling. In addition, germination was found to increase the value of buckwheat as a foodstuff because the amounts of GABA, minerals, amino acids, and other beneficial nutrients increased. This implies that health-conscious people would prefer to eat food made from germinated grains due to their increased nutrient content, such as buckwheat *natto* and *miso* paste. *Natto* and *miso* paste are traditional foods in Japan, and *natto* and *miso* paste have also become international products, preferred and consumed by many foreign people as well. Furthermore, the results of this study indicate that the germination treatment decreased the allergenic protein content in the buckwheat grains, allowing the production of foods with low- or no-allergenicity. Consequently, germinating buckwheat improves its properties, leading to a higher nutritional value, better taste and decreased allergenicity. In the future, we expect that germinated-buckwheat *natto* and *miso* paste will be consumed daily as healthy food choices.

REFERENCES

Handoyo, T., Maeda, T., Urisu, A., Adachi, T., & Morita. N. (2006). Hypoallergenic buckwheat flour preparation by *Rhizopus oligosporus* and its application to soba noodle. *Food Research International*, 39, 598-605.

Hung, P.V., & Morita, N. (2008). Distribution of phenolic compounds in the graded flours milled from whole buckwheat grains and their antioxidant capacities. *Food Chemistry,* 109(2) 325-331.

Kayashita, J., Shimaoka, I., Nakajoh, M., Yamazaki, M, & Kato, N. (1997). Consumption of buckwheat protein lowers plasma cholesterol and raises fecal neutral sterols in cholesterol-fed rats because of its low digestibility. *Journal of Nutrition*, 127(7), 1395-1400.

Kikunaga, S., & Takahashi, M. (1992). Biochemical changes in phosphorus compounds and in the activity of phytase and amylase in the buckwheat (*Fagopyrum esculentum)* grain during germination. *Bulletin of Notre Dame Seishin University*, 16, 61-64 (*in Japanese*).

Kobayashi, M., Hashimoto, Y., Taniuchi, S., & Tanabe, S. (2004). Degradation of wheat allergen in Japanese soy sauce. *International Journal of Molecular Medicine*, 13, 821-827.

Latta, M., & Eskin, M. (1980). A simple and rapid colorimetric method for phytate determination. *Journal of Agricultural and Food Chemistry*, 28, 1313-1315.

Miyake, K., Maeda, T., & Morita, N. (2006). Characteristics of germinated buckwheat and its application to the processing of buckwheat natto and *miso* paste. *Fagopyrum*, 23, 75-82.

Miyake, K., Morita, R., Handoyo, T., Maeda, T., & Morita, N. (2004). Characterizsation of graded buckwheat flours and some propeties of germinated 'Mancan' buckwheat grains. *Fagopyrum*, 21, 91-97.

Naka, T., Tujimoto, K., Gotou, T., Suekane, S., Asaga, M., Miyake, K., & Morita, N. (2011). Composition changes of rice-Tartary buckwheat derived from germination (p.98 (3Fa2)). Japanese Society for Food Technology, Abstracts of the 58th annual meeting, Sendai, Japan.

Oomah, B.D., & Mazza, G. (1996). Flavonoids and antioxidative activities in buckwheat. *Journal of Agricultural and Food Chemistry*, 44, 1746-1750.

Pomeranz, Y. (1983). Buckwheat: structure, composition, and utilization. *Critical Reviews in Food Science and Nutrition*, 19, 213-258.

Satake T. (2007). The Development of Process Technology for Highly Functional Rice. Japanese Society of Tasty Technology, 96 p.

Satake, T., Fukumori, T., Kanemoto, S., Kawano, M., Sasaki, Y., Liu, H., Ishiwata, K., Aoto, H., Shinmura, H., & Nakagawa, K. (2003). GABA generation by fine moistening, 4th Meeting of Japan Society for Food Engineering Abstract, p. 46.

Skrabanja, V., Liljeberg, H.G.M., Elmsttahl, X., Creft, I., & Bjorck, M.E. (2001). Nutritional properties of starch in buckwheat products: Studies in vitro and in vivo. *Journal of Agricultural and Food Chemistry,* 49, 490-496.

Srinivasan, K., Kaul, C.L., & Ramarao, P. (2005). Partial protective effect of rutin on multiple low dose streptozotocin-induced diabetes in mice. *Indian Journal of Pharmacology*, 37(5), 327-328.

Tomotake, H., Shimaoka, I., Katashita, J., Yokoyama, F., Nakajoh, M., & Kato, M. (2000), A buckwheat protein product suppresses gallstone formation and plasma cholesterol more strongly than soy protein isolate in hamster. *Journal of Nutrition*, 130, 1670-1674.

Undeger, U., Aydin, S., Basaran, A.A., & Basaran, N. (2004). The modulating effect of quercetin and rutin on the mitomysin C induced DNA damage. *Toxicology Letters*, 151, 43-49.

SECTION 3: PSEUDOCEREAL-BASED FUNCTIONAL FOOD DEVELOPMENT

In: Advances in Cereal and Pseudocereal ... ISBN: 978-1-62618-347-6
Editors: N. Morita, P.V. Hung et al.

Chapter 6

GERMINATED QUINOA FOR FUNCTIONAL FOODS

Naofumi Morita[1], Sang Ha Park[2] and Tomoko Maeda[3]

[1]Department of Food Packaging Technology, Toyo College of Food Technology, Mimami-Hanayashiki, Kawanishi, Hyogo, Japan

[2]Department of Food Science & Nutrition, College of Science and Engineering, Kyungpook National University, Sangju, Kyungpook, Korea

[3]Department of Science, Technology, and Human Life, Graduate School of Education, Hyogo University of Teacher Education, Hyogo, Japan

ABSTRACT

Quinoa is currently receiving a great deal of attention as a replacement for popular cereals with wheat flours to improve diet, as quinoa contains more nutrients and minerals than wheat flours. The flour characteristics and processing properties of quinoa for baked products have been presumed to be poor compared with wheat flours, making the incorporation of quinoa into foodstuffs difficult. However, using quinoa as a substitute for wheat flours may provide an advantage in reducing wheat-allergenic reactions. Therefore, to further develop the functional properties of quinoa flours, the grain was germinated. Quinoa germinated with a yield of 98% at 30°C for 48 hr. During the germination, total and essential amino acids increased 2.5 and 3.6-fold, respectively. In particular, GABA content and essential amino acid contents increased rapidly during the 24 hr-germination. Substitution of 10% germinated quinoa flour for wheat flour made distinctly harder dough. In a

Fermograph, the control and 24 hr-samples contained high amounts of both inner and total CO_2 in the gas cell. The loaf volume of the bread made with 24 hr-germinated quinoa flour substituted for wheat flour was the largest among the germinated samples; however, these differences were not statistically significant. Breadcrumbs of the control and 24 hr-samples became firm slower than those of the 48 and 72 hr-samples, indicating slower staling.

Linolenic acid (18:3) was the major glycolipid (GL) fatty acid in the control quinoa, and linoleic (18:2), oleic (18:1) and palmitic (16:0) acids were the major nonpolar lipid (NL), GL and phospholipid (PL) fatty acids in the germinated samples, respectively. During germination, oleic acid increased but linoleic acid decreased. The ratio of saturated, monounsaturated and polyunsaturated NL and PL fatty acids approached 3:4:3. Additionally, the ω3/ω6 ratio in NL, PL and GL became close to the value for unsaturated fatty acids in the traditional Japanese diet.

1. Introduction

In the history of mankind approximately 3,000 species of edible plants have been cultivated, but most of these gradually disappeared, with only 30 species currently cultivated. Quinoa is one of the cereals that have been cultivated in the Andean highlands (between 2,000～4,000 m above sea level) since 3,000 BC. Approximately 500 years ago, the Spanish conquest brought wheat and barley, foreign crops, which were alternated in most of the fields previously occupied by quinoa (Galway, 1990). Vietmeyer (1986) reported in the book "Amaranth: Modern Prospects for an Ancient crop" (Academy Press, 1984) that before the time of Cortez, grain amaranths were staple foods of the Aztec and Inca. However, quinoa was brought into the limelight by NASA's Controlled Ecological Life Support System because quinoa can reduce the CO_2 concentration in atmospheres in space and provide oxygen and various nutritional materials (Schlick, 1996). Quinoa is of special interest because of its nutritional value, especially its high protein and mineral contents. In addition, its essential amino acids make it suitable for a balanced human diet (Kozioł, 1992; Chauhan, 1992). The mineral content of quinoa has also been found to be greater than those of rice, wheat, buckwheat and maize grain crops. Conventionally, the leaves and sprouts can be eaten raw or cooked and provide a substantial amount of nutritive value (Johnson & Ward, 1993). Furthermore, the protein in quinoa flour causes no allergenic reactions and is biologically active (Konishi et al, 1995). For this reason, quinoa is under consideration for the commercial production of non-allergenic foods, especially for infants, in which wheat flour is substituted with amaranth flour. Quinoa as a food resource may provide greater nutrition than the major vegetative grain crops. In many grains, various nutrients and phytochemicals are present, mostly at the outer layer of the cereal grain, bran or seed coat. However, milling techniques typically remove these nutrient rich materials and they are used for animal feed. Therefore, we consume less nutritive cereals from polished rice grains or white breads

Recently, health conscious consumers in developed countries have begun to accept the consumption of coarse cereal, whole cereal flour, and less purified flour. Epidemiological studies strongly suggest that grain consumption plays a significant role in the prevention of many chronic diseases and has been associated with reduced risk of high blood pressure and cancer, a finding that has been associated in part with the phytochemicals in grains. These phenolic compounds such as ferulic acid, vanillic acid, caffeic acid, and p-coumaric acid are utilized as sources of nutritional antioxidants. Major portions of phenolics are located at the outer layer of grains in free or bound forms, except for the case of buckwheat. These bound phenolics are considered to be cleaved or hydrolyzed during the germination or sprouting of grains.

As already mentioned in this book, germination is the beginning of the life activity of seed grains, therefore the functional and nutritional properties of cereals are enhanced during this period (Chen, 1975; Wang & Fields, 1978; Aluko & Monu, 2003; Park & Morita, 2004; Park & Morita, 2005; Takao et al., 2005; Asao & Watanabe, 2010; Vega-Gálvez et al, 2010). In general, enzymatic activity in germination induces changes in major seed reserves such as carbohydrate, lipid and protein reserves. Consequently, germination leads to an increase in free amino acids, mono- and diacylglycerol, free fatty acids, sugar, vitamins, phenolics, phytic acid and minerals (Vanderstoep, 1981). In addition, during germination, the content of anti-nutritional materials such as trypsin inhibitors (El-Mahdy, 1985), tannins (Udayasekhara, 1995), pentosans and phytic acid is reduced (Vidal-Valverde, 1994). Hence, the germination of quinoa seeds is considered beneficial from the viewpoint of nutritional value. Unfortunately, along with the development of milling techniques, these nutrient rich materials are normally removed and used as animal food.

The purpose of this chapter is to evaluate the functional changes that occur during germination in quinoa seeds and to explore the application of quinoa flour for food processing.

2. Experimental Methods

Quinoa seeds, *Chenopodium quinoa* Willd, were obtained from Dai-Nippon Meiji Sugar Co., Ltd., (Sakai, Japan) following importation from Peru. Wheat flour used to make bread was the commercial hard wheat flour "Cameria", obtained from Nisshin Flour Milling, Inc. (Kobe, Japan).

Germination of quinoa seeds (300 g) in distilled water (2 L) was carried out for 24, 48 and 72 hr using an electric germination machine (HP-70, Takegosi Seisakusho Co., Ltd., Tubame, Japan) at 30°C. During incubation, quinoa seeds were rinsed twice a day with 400 ml of distilled water. After incubation, samples were washed carefully with distilled water and then freeze-dried after freezing with liquid nitrogen. Freeze-dried samples were ground to pass through a 60-mesh sieve using an analytical mill.

The viscoelastic properties of the dough after mixing in the farinograph were measured with a Fudoh rheometer (RT-2002D·D, Rheotec Co., Ltd., Tokyo, Japan) as previously reported (Morita et al., 1999).

2.1. Bread Baking

The breadmaking procedures were performed following a slightly modified AACC method (10-10B 1995). The ingredients flour (300g), sugar (15g) and salt (4.5g) were mixed in a polyethylene bag (Park & Morita, 2005). Four grams of dry baker's yeast (Asahi Kasei Co., Ltd., Tokyo, Japan) and an appropriate amount of water were added and mixed using a Ladies mixer KN-200 (Taisho Denki Co., Ltd., Osaka, Japan) for 15 min at a speed of 380 rpm. The optimum amount of water was determined from the water absorption using the farinograph at 500 BU. The mixed dough was processed by three-step fermentation. The first and second fermentations were carried out at 30°C and 85% relative humidity for 60 min each. After the second fermentation, the dough was divided into three pieces (130g), rounded and molded. The molding was created using an SM-230 molder (Baker's Production Co., Ltd., Osaka). Then, the final fermentation was carried out for 30 min at 38°C and 90% relative humidity. The bread was baked in a POF-1 oven (Baker's Production Co., Ltd.) for 20 min at 200°C. The loaf volumes were measured using the rapeseed displacement method.

2.2. Breadcrumb Firmness

The breadcrumb firmness was assessed using a rheometer (Rheotec Co., Ltd., Tokyo, Japan), as reported previously (Morita et al., 2001). The plunger was penetrated into the sample at 6 cm/min, and the data were processed using the computer program Rheosoft TR-06 (Rheotech Co., Ltd.).

2.3. Fractionation of Proteins and Analysis of Free Amino Acids

Flour protein was fractionated by a procedure modified from Osborne (1907). Five grams of flour were extracted stepwise with water, 5% NaCl, 70% ethanol and 0.2% NaOH aqueous solution. After extraction, the mixtures were centrifuged at 1,200×*g* for 15 min and nitrogen content was determined in each of the supernatants and residues using the micro-Kjeldahl method. Free amino acids were determined using a Hitachi L-8,500 amino acid analyzer. Each sample (5g) of germinated flour was extracted with 15 ml of 75% ethanol aqueous solution in triplicate. The extracts were combined and evaporated until dry. The residue was filled up to 25 ml with 0.02 N HCl and filtered through a 0.45 μm membrane filter. The filtrates were analyzed for free amino acids using a 25 cm stainless steel column prepacked with cation exchange resin. Tryptophan content was not measured.

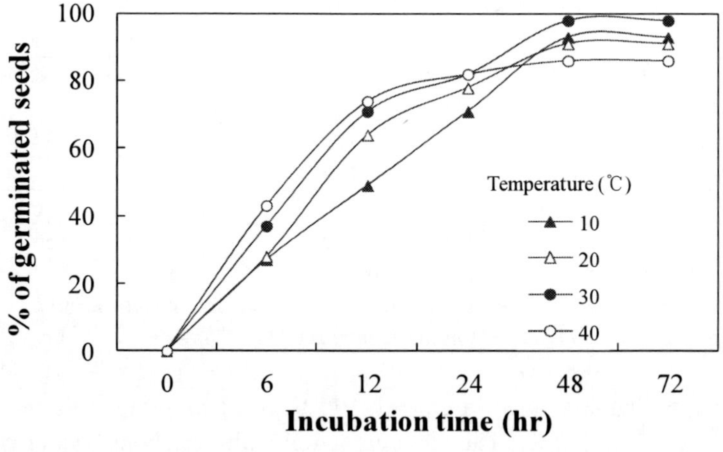

Figure 6-1. Time course of the germination of quinoa seeds.

3. Experimental Results and Discussion

3.1. The Germination Rate

The effect of temperature on the germination rate of quinoa seeds was tested at 10, 20, 30 and 40°C. After 24 hr germination, approximately 81% of the quinoa seeds germinated, and the germination rate increased to 98% after 48 hr at 30°C (Fig. 6-1). At 72 hr, the highest germination rate was 30°C, followed by 20, 10 and 40°C.

3.2. Fractionation of Proteins and Free Amino Acid Analyses

Table 6-1 shows the amounts of various proteins fractionated from ungerminated, 24, 48 and 72 hr-germinated quinoa flours, including albumin, globulin, prolamin, glutelin and residue (alkaline-insoluble protein). In the ungerminated quinoa flour, the most abundant protein was albumin, followed by alkaline-insoluble protein, glutelin, globulin, and prolamin; whereas the most abundant protein in the 72 hr germinated quinoa flour was alkaline-insoluble protein, followed by glutelin, globulin, albumin and prolamin.

Table 6-1. Protein fractionation of ungerminated (control, 0), 24, 48 and 72 hr-germinated quinoa flours, expressed as a percentage of extracted protein* (%, w/w)

Germination time (hr)	Albumins (%)	Globulins (%)	Prolamins (%)	Glutelins (%)	Residues (%)
0 (control)	31.6^{c}	30.3^{b}	2.7^{c}	4.0^{a}	31.4^{a}
24	16.9^{b}	25.9^{b}	0.8^{b}	15.6^{b}	39.8^{a}
48	14.5^{b}	15.2^{a}	0.5^{ab}	19.1^{c}	50.3^{b}
72	10.8^{a}	12.5^{a}	0.2^{a}	20.8^{c}	55.5^{b}

*Values are expressed on a dry weight basis.
Values followed by the same letter in the same column are not significantly. different according to Duncan's multiple range test ($P<0.05$); n=4.

The prolamin content was only 0.2% of total protein in the 72 hr-germinated quinoa flour. The albumin and globulin fractions of ungerminated quinoa flour decreased slightly during germination. The glutelin fraction increased significantly with germination. In the 72 hr-germinated quinoa flour, large amounts of insoluble proteins might be strongly correlated with baking quality, as reported previously (Van den Berg & Bruin, 1981).

Table 6-2 shows the free amino acid content of quinoa flour at different germination times. After a 48 hr-germination, the total and essential free amino acid contents were 2.5 and 3.6-fold of the levels in ungerminated quinoa flour, respectively. In particular, gamma-aminobutyric acid (GABA) content increased rapidly during the first 24 hr of germination and then steadily decreased after 48 hr of germination. These changes in the fractionated protein and essential free amino acid contents of germinated quinoa were assumed to result from the degradation and denaturation of protein by the reaction with protease throughout the germination period to supply the growth of plant cells.

Table 6-2. Changes in free amino acid composition during the germination of quinoa seeds (mg/100g, w/w)

Amino acids	**Incubation time (hr)**			
	0	**24**	**48**	**72**
Thr	4.2[a]	24.7[cd]	27.3[d]	21.7[c]
Val	13.0[a]	37.5[c]	48.0[d]	35.3[c]
Met	3.7[a]	14.6[c]	23.3[e]	19.2[d]
Leu	6.5[a]	51.2[c]	68.3[d]	66.9[d]
Phe	8.3[a]	36.8[c]	58.6[e]	49.0[d]
I-leu	7.4[a]	28.3[c]	36.8[d]	27.1[c]
Lys	7.6[a]	28.5[b]	40.1[c]	29.7[b]
His	39.2[c]	36.8[c]	19.4[b]	8.5[a]
Gly	6.2[a]	22.6[c]	21.9[c]	14.2[b]
Ala	23.2[a]	44.8[c]	45.5[c]	32.6[b]
Ser	6.7[a]	23.3[d]	22.4[d]	18.8[c]
Cys	---*	2.2[b]	4.2[d]	3.4[c]
Pro	7.5[a]	32.2[c]	50.0[d]	46.4[d]
Thr	16.2[a]	49.6[c]	56.2[d]	50.3[c]
GLU**	19.8[a]	78.9[d]	59.6[c]	47.3[b]
ASP***	31.4[d]	14.6[b]	19.8[c]	20.5[c]
Arg	74.9[c]	84.9[d]	69.3[b]	52.6[a]
GABA	9.6[a]	59.6[d]	48.4[c]	27.4[b]
Total	**285.3**	**671.2**	**719.1**	**570.8**

Values followed by the same letter in the same row are not significantly different according to Duncan's multiple range test ($P<0.05$); n=4.

* Trace values were not detected. ** Glutamic acid + glutamine. *** Aspartic acid + asparagines.

3.3. Thermal Transitions of Starch

Table 6-3 shows the gelatinization properties of ungerminated quinoa flour and 24-, 48- and 72 hr-germinated quinoa flour samples using DSC. The endothermic peaks of all samples were observed along with gelatinization approximately 44.8 - 72.8°C. The ranges of thermal transitions were similar among all samples; onset (*T*o), peak (*T*p) and conclusion (*T*c) temperatures were observed for all samples, and the thermal characteristics of these samples do not appear to be significantly different. However, the ΔH of 24-, 48- and 72 hr-germinated quinoa flour was larger than that of ungerminated quinoa flour (1.32, 1.21 and 1.14-fold larger, respectively). This may be because during the early stage of germination, activated α-amylase and lipase hydrolyze amylopectin and lipid, respectively. Thus, the liberated fatty acid combined with the smaller molecular sizes of amylopectin.

Table 6-3. Gelatinization characteristics of ungerminated (0) and 24, 48 and 72 hr-germinated quinoa starches

Incubation time (hr)	*T*o (°C)	*T*p (°C)	*T*c(°C)	ΔH (J/g)
0	45.1±0.7	58.1±0.3	72.1±1.3	10.5±1.0
24	46.1±0.5	56.6±0.6	72.8±1.4	13.9±1.5
48	44.8±1.1	58.2±1.4	71.3±1.3	12.7±1.8
72	45.5±1.0	56.9±0.2	72.6±2.0	12.0±0.4

Abbreviations: *T*o, onset temperature; *T*p, peak temperature; *T*c, conclusion temperature; ΔH, gelatinization enthalpy change of starch (J/g); n=6.

3.4. Viscoelastic Properties

Wheat flour was substituted with 10% control (0), 24, 48 or 72 hr-germinated quinoa flour and the effects on the viscoelastic properties of the dough were evaluated (Table 6-4). The compression stress, modulus of elasticity, and viscosity coefficient of the control dough were significantly lower than those of the wheat dough substituted with 24, 48 and 72 hr-germinated quinoa flour. Among all the tested samples, the 24 hr sample showed the highest mixing parameters, but no significant differences were observed in the viscoelastic properties among the 24, 48 and 72 hr-germinated quinoa flour wheat dough samples. As for the 48 and 72 hr-germinated dough samples, the viscoelastic properties and farinograph data were higher than those of the sound flour.

Table 6-4. Viscoelasticity of wheat dough containing 10% of control (0), 24, 48 and 72 hr-germinated quinoa flour

Incubation time (hr)	Stress (kPa)	Elastic modulus (kPa)	Viscosity coefficient (10^2kPa·)
Control(0)	15.7±2.8[a]	47.3±4.9[a]	1.2±0.2[a]
24	26.9±3.5[b]	80.5±6.3[b]	2.0±0.4[b]
48	24.5±1.9[b]	76.7±3.4[b]	2.0±0.4[b]
72	23.7±2.6[b]	72.7±5.6[b]	1.9±0.3[b]

Values followed by the same letter in the same column are not significantly different according to Duncan's multiple range test (P<0.05). Average±standard deviation, n=4 (adapted from Park & Morita, 2005).

3.5. Fermograph Data

To determine the gas retention of the wheat dough substituted with sound (0), 24, 48 and 72 hr-germinated quinoa flour during dough fermentation, the total and inner gas generations were measured as shown in Figures 6-2A and 6-2B. The results of the total and inner gas generation during 120 min of fermentation were similar among all samples. However, two patterns were observed from 120 min in Figure 6-2A and from 140 min in Figure 6-2B, respectively. The amounts of total and inner gas generated by the 48 and 72 hr-samples were lower than those of the sound and 24 h-samples, with the amount of inner gas generated after 200 min equal to 91% of that generated by the sound and 24-hr samples. These results suggest that the wheat dough samples with control (sound) and 24 hr-germinated quinoa flour have more extensible gluten matrixes than the 48 and 72 hr-samples. Moreover, these results were consistent with the specific volume of bread, as shown in Figure 6-3.

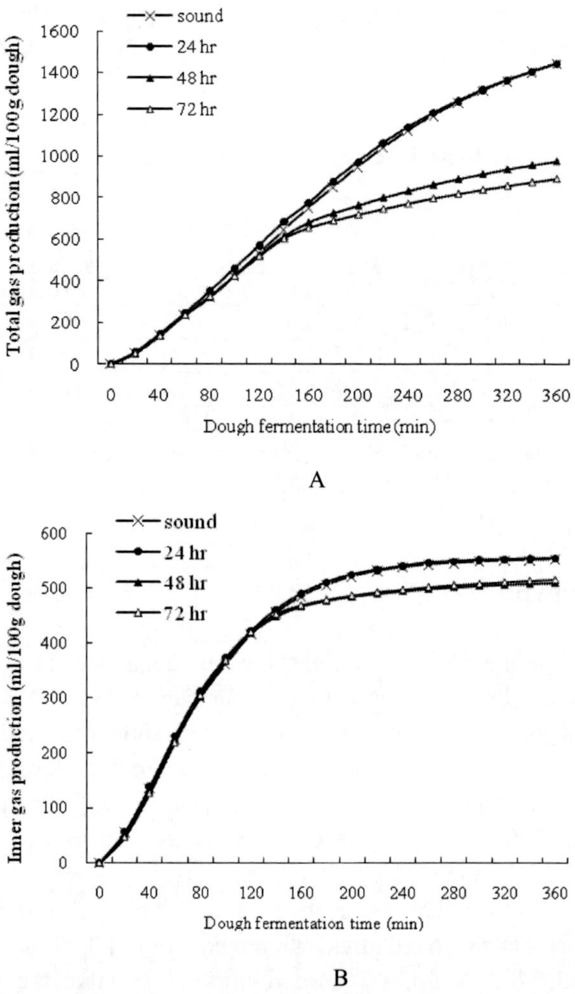

Figure 6-2. Total (A) and inner (B) gas generation of wheat dough containing 10% of sound (0), 24, 48 and 72 hr-germinated quinoa flour (adapted from Park & Morita, 2005).

3.6. Breadmaking Results

Figure 6-3 shows the baking results with sound (0), 24, 48 and 72 hr-germinated quinoa flour substituted for wheat flour. The loaf volume of the sound was significantly larger than those of the 48 and 72 h-samples, while that of the 24 hr-sample was slightly larger than that of the sound. These results were also expected based on the farinograph, SEM and fermogram results during the dough mixing period. The weak gluten matrix as shown by the low valorimeter value resulted in a discontinuous gluten structure and of the generation of little gas in the dough. Therefore, the breadcrumb was not able to maintain sufficient gas during breadmaking (Baker et al., 1945; Macritchie, 1976). Moreover, the results after storage of the breadcrumbs made from the sound (0), 24, 48 and 72 hr-samples coincided with those of the breadmaking results as shown in Figure 6-4. The staling of the bread was also studied, and breadcrumbs of the control and 24 hr samples became firm more slowly than the 48 and 72 hr-samples.

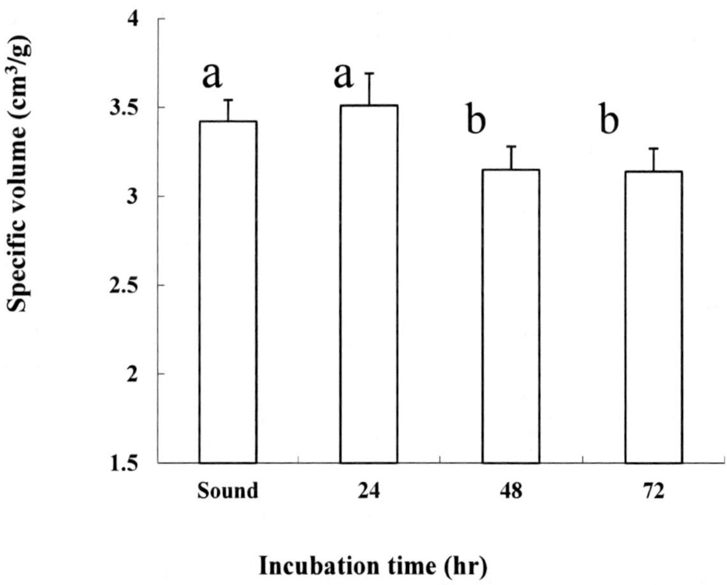

Figure 6-3. Effect of substitution with 10% of sound (0), 24, 48 and 72 hr-germinated quinoa flour on the specific volume of bread. Abbreviations used here are the same as those in Table 6-1. Bars labeled with the same letter are not significantly different according to Duncan's multiple range test (P<0.05). Average± standard deviation, n=4..

3.7. Changes in Bound Lipids and the Fatty Acid Composition during the Germination of Quinoa Seeds

Quinoa whole grains contain approximately 7~8% total lipids, and these lipids make distinct contributions to improving the quality of the diet because of the presence of a large amount of essential fatty acids such as linoleic and linolenic acids (Morita et al., 2001). Many studies have reported on the lipid content and fatty acid composition of ether extractable materials of quinoa (Wood et al., 1993; Ando et al., 2002). However, little research has been carried out on the changes in bound lipids (BL) and total lipids (TL) during the germination of quinoa. Triacylglycerols (TAGs), a major portion of the lipids stored in seeds, were catalyzed by lipases to yield both glycerol and free fatty acids (FFA). Next, the products of lipid degradation, glycerol and FFA, are usually used again in the synthesis of TAGs. This metabolic pathway induces changes in the ω3/ω6 ratio and in the saturated:monosaturated:polysaturated fatty acid composition. For this reason, the free and bound lipids formed during the germination of quinoa seeds could be expected to induce new functionality and nutritional value.

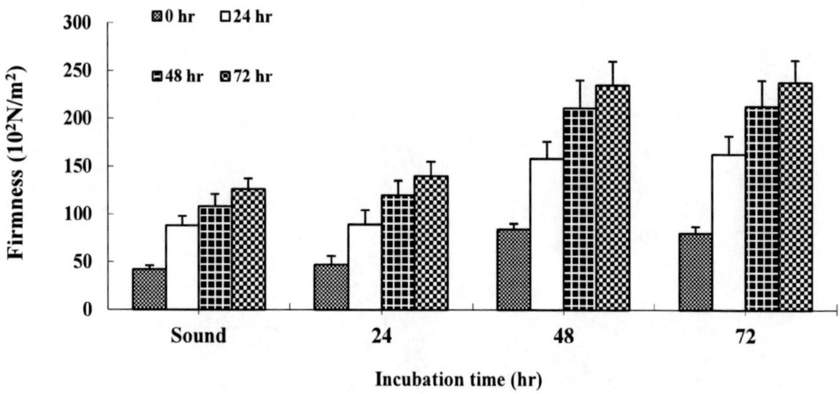

Figure 6-4. Staling of breadcrumbs containing 10% of sound (0), 24, 48 and 72 hr-germinated quinoa flour during 3 days of storage. Abbreviations used here are the same as those in Table 6-1. Bars labeled with the same letter are not significantly different according to Duncan's multiple range test ($P<0.05$). Average±standard deviation, n=4.

3.8. Fatty Acid Compositions of NL (Nonpolar Lipids)

The major NL fatty acids present during the germination of quinoa were oleic acid, linoleic acid and palmitic acid, as shown in Table 6-5. In NL, oleic acid significantly increased after 72 hr of germination compared to the control, 24, and 48 hr-seeds. In contrast, linoleic acid significantly decreased after 72 hr of germination compared to the control, 24, and 48 hr-samples. A large amount of arachidonic acid (20:0) was observed in 48 hr-germinated quinoa. The amount of palmitic acid significantly after 72 hr of germination decreased compared to the control, 24 and 48 hr-germinated quinoa flours. After 72 hr of germination, the proportions of saturated, monounsaturated and polyunsaturated fatty acids were improved to approach 3:4:3. However, the ratio of polyunsaturated to saturated fatty acids decreased during the germination.

3.9. Fatty Acid Composition of PL (Phospholipids)

Many germinated samples contained linoleic acid as the most predominant fatty acid, except for 72 hr-germinated samples, while the sample germinated for 72 hr contained oleic acid as the most predominant fatty acid (42.4%), with linoleic and palmitic acids present in similar amounts, as shown in Table 6-6. From these results above, the linolenic acid was the highest in GL (glycolipids), followed by PL and NL.

Table 6-5. Fatty acid compositions of nonpolar lipids

Fatty acid compositions		**Incubation time (hr)**			
		Control	**24**	**48**	72
14:0	Myristic	0.7±0.1[b]	0.2±0.0[a]	0.2±0.0[a]	0.6±0.1[b]
16:0	Palmitic	21.9±2.4[c]	16.3±2.4[b]	10.4±1.2[a]	19.6±1.3[bc]
18:0	Stearic	2.3±0.2[b]	2.7±0.3[b]	0.9±0.2[a]	2.1±0.6[b]
18:1	Oleic	22.0±2.1[a]	26.4±2.8[ab]	30.6±0.8[b]	40.7±0.9[c]
18:2	Linoleic	49.1±3.5[b]	42.1±4.8[b]	47.0±2.8[b]	26.9±3.7[a]
18:3	Linolenic	3.4±0.1[b]	0.6±0.3[a]	0.5±0.3[a]	2.9±0.6[b]
20:0	Arachidic	0.1±0.0[a]	3.8±0.3[c]	7.8±0.3[d]	0.9±0.2[b]
Saturated		25.1	25.0	19.8	24.8
Monounsaturated		22.1	28.7	31.4	43.4
Polyunsaturated		52.8	46.3	48.8	31.8
ω3/ω6		0.07	0.01	0.01	0.11

Each value shows the percentage as the dry weight basis.

Values followed by the same letter in the same row are not significantly different according to Duncan's multiple range test ($P<0.05$); n=3 (adapted from Park & Morita, 2004).

Table 6-6. Fatty acid compositions of phospholipids

Fatty acid compositions		Incubation time (hr)			
		Control	24	48	72
14:0	Myristic	0.2±0.0[a]	0.2±0.0[a]	1.7±0.1[c]	0.6±0.1[b]
16:0	Palmitic	16.4±1.6[a]	17.9±2.3[a]	14.8±1.5[a]	25.4±1.1[b]
18:0	Stearic	0.7±0.2[a]	1.0±0.2[a]	1.3±0.3[ab]	2.1±0.5[b]
18:1	Oleic	23.4±1.9[a]	24.5±1.9[a]	27.0±2.8[a]	42.4±5.3[b]
18:2	Linoleic	54.6±3.6[b]	53.6±2.8[b]	52.2±3.1[b]	26.0±0.9[a]
18:3	Linolenic	4.3±0.3[c]	2.8±0.4[ab]	3.0±0.4[b]	2.3±0.2[a]
Saturated		17.4	19.1	17.8	29.1
Monounsaturated		23.5	24.5	27.0	42.5
Polyunsaturated		59.0	56.4	55.2	28.4
ω3/ω6		0.08	0.05	0.06	0.09

Each value shows the percentage as the dry weight basis.

Values followed by the same letter in the same row are not significantly different according to Duncan's multiple range test (P<0.05); n=3 (adapted from Park & Morita, 2004).

After germination for 72 hr, the ratio of saturated, monounsaturated and polyunsaturated fatty acids of NL and PL became 3:4:3, excluding GL. Additionally, the ω3/ω6 ratio in NL, PL and GL became close to the value of unsaturated fatty acids in the traditional Japanese diet. After 72 hr of germination, linoleic acid significantly decreased in NL, GL and PL, while oleic acid increased. The reason for the increase in the amount of oleic acid along with germination remains unknown. Although linoleic acid content decreased after 72 hr of quinoa germination, this would not significantly decrease the total amount of linoleic acid in the substituted foodstuff because the quinoa is substituted in a small amount. From the viewpoint of nutrition, germinated quinoa flour is expected to increase functional components similar to the commercially available brown rice and buckwheat flour.

CONCLUSION

Whole quinoa grain contains various functional and nutritional materials, as it belongs to the family of dicotyledonous plants rather than monocotyledonous plants such as poaceae. Additionally, quinoa grain has rarely been incorporated as part of our daily diet, and allergenic reactions have not been reported (Konishi, 2002) to the best of my knowledge. Furthermore, germination of grain is a suitable method to its increase functional and nutritional properties. During the germination, total and essential amino acids increased 2.5 and 3.6-fold, respectively, compared to ungerminated flour. In particular, GABA content increased rapidly during the first 24 hr of germination and then decreased after 48 hr of germination.

Substitution of wheat dough with 10% germinated quinoa flour decreased the peak viscosity values of the amylograph in proportion to germination time. This result could be due to the activity of enzymes during germination such as amylase and protease. Thus, the control and 24 hr-samples contained high amounts of both inner and total CO_2 in the gas cell. This result has a close relationship to loaf volume, one of the most important qualities for breadmaking, as loaf volume generally depends on the amount of gas generated during dough fermentation. Consequently, the bread prepared with 24 hr-germinated quinoa flour had the largest loaf volume, followed by the control, 48 and 72 hr-samples. Thus, from the viewpoint of nutrition and bread quality, the increased levels of free amino acids and the large loaf volume make the 24 hr-germinated quinoa flour a useful substitute for practical preparation of nutritional bread.

In conclusion, the partial substitution of quinoa flour for "Cameria" reduced baking quality, such as the staling of breadcrumbs, the specific volume, and the physical properties of the dough. However, the reduced baking quality could be improved by the use of fungal lipase and hemicellulase (data not shown), which can improve dough fermentation. In addition, germination was considered to benefit human diets by increasing the nutritional value of the seeds. Thus, this study can further the development of quinoa as a useful ingredient in food processing.

REFERENCES

Aluko, R.E., & Monu, E. (2003). Functional and bioactive properties of quinoa seed protein hydrolysates. *Journal of Food Science*, 68, 1254-1258.

Asao, M., & Watanabe, K. (2010). Functional and bioactive properties of quinoa and amaranth. *Food Science and Technology Research*, 16, 163-168.

Ando, H., Chen, Y. C., Tang, H., Shimizu, M., Watanabe, K., & Mitsunaga, T. (2002). Food components in fractions of quinoa seed. *Food Science and Technology Research*, 8, 80-84.

Chauhan, G. S., Eskin, N. A. M. & Tkachuk, R. (1992). Nutrients & antinutrients in quinoa seed. *Cereal Chemistry,* 69, 85-88.

Chen, N. H., Wells, C. E. & Fordham, J. R. (1975). Germinated seeds for human consumption. *Journal of Food Science*, 40, 1290-1295.

El-Mahdy A. R., Moharram Y. G., & Abou-Samaha, O. R. (1985). Influence of germination on the nutritional quality of lentil seeds. *Zeitschrift für Lebensmittel Untersuchung und Forschung* , 181, 318-320.

Galway, N. W., Leakey, C. L. A., Price, K. P., & Fenwick, G. R. (1990). Chemical composition and nutritional characteristics of quinoa (*Chenopodium quinoa Willd.*). *Journal of Food Sciences and Nutrition,* 4, 245-261.

Johnson, D. L., & Ward, S. M. (1993). Quinoa. In: Janick, J., & Simon, J. E. (Eds), *New Crops*. pp. 219-221. Wiley, New York.

Konishi, Y. (2002). Nutritional characteristics of pseudocereal amaranth and quinoa: Alternative foodstuff for patients with food allergy. *Journal of Japan Society of Nutrition and Food Sciences,* 55, 299-302. (in Japanese)

Konishi,Y. Nakamura, R., Sugimoto Y., & Fujimori, K. (1995). Globulin and Albumin-2 Associated with Protein Bodies in Amaranthus cruentus Seeds. *Bioscience, Biotechnology, and Biochemistry*, 59, 2140-2142.

Kozioł, M. J. (1992). Chemical composition and nutritional evaluation of quinoa (*chenopodium quinoa Willd.*). *Journal of Food Composition and Analysis,* 5, 35-68.

Morita N., Kang W., Hamauzu Z., & Sugimoto Y. (1999). Effect of amaranth flour on some properties of wheat dough and bread. *Journal of Applied Glycoscience*, 46, 23-30.

Morita N., Hirata C., Park S. H., & Mitsunaga T. (2001). Quinoa flour as a new foodstuff for improving dough and bread. *Journal of Applied Glycoscience*, 48, 263-270.

Osborne, T.B. (1907). The proteins of wheat kernel. Carnegie Inst., Washington, DC.

Park, S.H., & Morita, N. (2004). Changes of bound lipids and composition of fatty acids in germination of quinoa seeds. *Food Science and Technology Research,* 10, 303-306.

Park, S.H., & Morita, N. (2005). Dough and breadmaking properties of wheat flour substituted by 10% with germinated quinoa flour. *Food Science and Technology Research*, 11, 471-476.

Park, S.H., Maeda, T., & Morita, N. (2005). Effect of whole quinoa flours and lipase on the chemical, rheological and breadmaking charateristics of wheat flour. *Journal of Applied Glycoscience*, 52, 337-343.

Schlick, G., & Bubenheim, D. L. (1996). Quinoa: candidate crop for NASA's controlled ecological life support systems. In: Janick, J. (Ed), *Progress in New Crops*. pp. 632-640. ASHS press, Arlington, VA.

Takao, T., Watanabe, N., Yuhara, K., Itoh, S., Suda, S., Tsuruoka, Y., Nakatsugawa, K., & Konishi, Y. (2005). Hypocholesterolemic effect of protein isolated from quinoa (*Chenopodium quinoa* Willd.) seeds. *Food Science and Technology Research*, 11, 161-167.

Udayasekhara, Rao. P. (1995). Effect of germination on tannin, mineral and trace element composition on groundnut varieties. *Journal of the American Oil Chemists' Society*, 72, 477-480.

Vidal-Valverde, C., Frias, J., Estrella, I., Gorospe, M. J., Ruiz, R., & Bacon, J. (1994). Effect of processing on some antinutritional factors of lentils. *Journal of Agricultural and Food Chemistry*, 42, 2291-2295.

Van den Berg, C., & Bruin, S. (1981). Water activity and its estimation in food systems: Theoretical aspects. In Rockland, L. B., & Stevart, G. F. (Eds), *Water Activity: Influences on Food Quality*. pp. 1-24. Acedemic Press, New York..

Vanderstoep, J. (1981). Effect of germination on the nutritive value of legumes. *Food Techhnology*, 3, 83-88.

Vega-Gálvez, A., Miranda, M., Vergara, J., Uribe, E., Puente, L., & Martínez, E.A. (2010). Nutrition facts and functional potential of quinoa (*Chenopodium quinoa* willd.), an ancient Andean grain: a review. *Journal of Science of Food and Agriculture*, 90, 2541-2547.

Vietmeyer, N. D. (1986). Lesser-known plants of potential use in agriculture and forestry. *Science*, 232, 1379-1384.

Wang, Y. D., & Fields, M. L. (1978). Germination of corn and sorghum in the home to improve nutritive value. *Journal of Food Science*, 43, 1113-1119.

Wood, S. G., Lawson, L. D., Fairbanks, D. J., Robisom, L. R., & Andersen, W. R. (1993).Seed lipids content and fatty acid composition of three quinoa cultivars. *Journal of Food Composition and Analysis*, 6, 41-44.

In: Advances in Cereal and Pseudocereal ... ISBN: 978-1-62618-347-6
Editors: N. Morita, P.V. Hung et al.

Chapter 7

GERMINATED AMARANTH FOR FUNCTIONAL FOODS

Naofumi Morita[1], Motoki Yamaoka[2] and Tomoko Maeda[3]

[1]Department of Food Packaging Technology, Toyo College of Food Technology, Mimami-Hanayashiki, Kawanishi, Hyogo, Japan

[2]R&D Department, Maruha Nichiro Seafoods, Inc., Chuo-ku, Sapporo, Hokkaido, Japan

[3]Department of Science, Technology, and Human Life, Graduate School of Education, Hyogo University of Teacher Education, Shimokume, Kato, Hyogo, Japan

ABSTRACT

Amaranth, a type of pseudo-cereal, has been gaining attention in food processing because of its nutritional benefits. In this chapter, two types of Amaranth grains (*Amaranthus hypochondriacus* and *A. cruentus*) were used for germination, and the changes in the nutritional composition and functional properties of the germinated grain were investigated. During germination, free amino acids, dietary fibers and anti-oxidant activities of the amaranth grains increased when the germination time increased. The properties of proteins were also changed by the higher activity of enzymes, and an increase in the amount of albumin and a decrease in the amounts of globulin and glutelin were observed. Compared to the control without germination, the immunoblotting reactions of proteins derived from germinated amaranth against the sera of patients with a wheat-allergen reaction were strongly positive; however, those positive reactions weakened during the germination process. As a result, the germinated amaranth, with its low-allergenic and high-functional properties, is considered to be a new substitute for wheat, a popular cereal.

INTRODUCTION

Amaranth, an annual dicotyledonous plant belonging to the Amaranth family, is classified into three categories of cultivar: *A. hypochondriacus*, cultivated in Mexico; *A. caudatus*, cultivated in Peru; and *A. cruentus*, cultivated in Guatemala. Amaranth is classified as a pseudo-cereal, similar to foxtail millet, milium and millet, because it contains a higher amount of protein than many of the major cereal grains, such as rice, wheat and corn (Kauffman, 1992). Furthermore, amaranth also has a high amino acid score and sufficient amounts of lipids and minerals. In general, rice and wheat flour contain low levels of lysine, an essential amino acid. However, amaranth is reported to contain two times the level of lysine compared to rice. As a result, amaranth has received much attention as an economically valuable, unused resource according to the National Academy of Science (1984). Today, this extremely nutritious food has a bright future. Furthermore, amaranth grain has a lowering effect on blood cholesterol levels (Ogawa et al., 2000) and curative properties against carbohydrate metabolism.

In recent years, according to the development of the western lifestyle, the modernization of dietary habits has progressed along with diversification and facilitation of consumer needs. At the same time, humans are facing issues such as overeating and full feeding, but the balance of nutritional value is prone to worsen. The incidence of three major diseases associated with adult lifestyle habits—cancer, cardiac disorder and cerebrovascular disease—has increased approximately 2.5-fold in the last 60 years. Epidemiological studies strongly suggest that grain consumption plays a significant role in the prevention of many chronic diseases and has been associated with a reduced risk of certain chronic conditions, such as high blood pressure and cancer; these effects have been partly associated with the phytochemicals in grains. These phenolic compounds include ferulic acid, vanillic acid, caffeic acid and p-coumaric acid, and they are utilized as a source of nutritional antioxidants. The major parts of these phenolics are located on the outer layer of grains in either the free or bound forms. The bound phenolics are cleaved or hydrolyzed during the germination or sprouting of grains.

Due to an increase in the awareness of food effects on our health, the intake of pseudo-cereal as part of our diet has steadily attracted attention (Morita et al., 1999; Park & Morita, 2004; Oszvald et al., 2009; Mlakar et al., 2009). Therefore, the study of valuable, unused plant resources, such as amaranth, is quite important.

To the best of our knowledge, there has been no extensive research thus far on the germination of amaranth similar to the existing studies on brown rice, wheat, buckwheat and soybean. The authors aimed to investigate amaranth germination with regard to the functional materials formed and the mineral components changed during germination and to compare these results with those for other grains, such as wheat and rice.

Grain and Germination

Two types of amaranth grains were used: A (*A. hypochondriacus,* New Azteca) was obtained from Iwate Prefecture (Japan), and B (*A. cruentus*) was obtained from Shinkyo Sangyo Co., Ltd. (Yamaguchi, Japan) as an import from Peru.

For the germination test, 100 grains were placed on Kleenex tissue in flat dishes. Then, enough water was added to barely cover the grain. The dishes were incubated at 10, 20, 30, 40 and 50°C, under 85% relative humidity and with or without fluorescent light at various time intervals. Germinated grains were washed with pure water and freeze-dried. Next, the germinated grains were pulverized to the size of 120 meshes and stored in a refrigerator with desiccant until use.

Adequate germination was not observed at 10°C and 50°C, but the germination of grain B under dark conditions began after 24 hr and reached 60% at 20°C after 72 hr. Both grains A and B germinated well at 30°C and reached approximately 90% germination under dark conditions, but under light conditions, grain B reached 80% germination and grain A reached 60% germination. At incubation temperature of 40°C, delayed germination was observed until 12 hr of incubation for both the A and B grains under light conditions, but after 12 hr, the germination reached more than 90%.

The dry weight contents of germinated grains A and B were both 56% of the original grain weight after 12 hr of incubation; after 48 hr, A was 28% and B was 12%; and after 72 hr, A was 20% and B was 8%.

Fluctuation of Various Enzymes during Germination

It is well known that the activity of various enzymes increases along with the germination of grains. Here, representative enzyme activities such as amylase, protease and lipase were tested. In this experiment, each enzyme was extracted from germinated grains A and B. Alpha-amylase activity reached a maximum (A: 1.5 U/mg and B: 1.0 U/mg protein for potato starch as the substrate; A: 1.3 U/mg and B: 1.0 U/mg protein for wheat starch as the substrate) after approximately 24 hr of incubation and then decreased gradually. The average chain length of polysaccharides in the powder was reduced from 10-12 U of the ungerminated value to approximately 1.45 U after 48 hr.

Protease activity also showed a maximum value after 24 hr of incubation (B: 0.015 U/mg; A: 0.01 U/mg protein), and the activity decreased more quickly than that of amylase. However, lipase activity increased quite slowly. After 24 hr of incubation, the activity had reached a low value (A and B: 0.05, where absorbance at 410 nm was due to p-nitrophenol), and the activity increased to values of 0.55 for grain B and 0.30 for grain A after 48 hr of incubation.

Change of Free Amino Acid Amounts in Amaranth Grains during the Germination Process

According to the length of the germination time, the amounts of free amino acids in the grains mostly increased (Table 7.1). After 72 hr of germination, the total amounts of free amino acids in the A and B grains were approximately 6 and 16 times higher, respectively, compared to the non-germinated grain (0 hr, the control). Therefore, the nutritional improvement was clearly a result of the longer germination time. In addition, the amounts of aspartic acid (15 and 62 times), glutamic acid (5 and 55 times), valine (7 and 13 times), leucine (8 and 15 times), isoleucine (7 and 13 times), lysine (13 and 24 times) and histidine (7 and 14 times) increased in the A and B grains; the values in parentheses are the relative volumes of A and B compared to the control. Ganma-aminobutyric acid (GABA), which can enhance blood-flow, extend the memory period and improve mental disorder (Sharpley, 2009), also increased approximately 5- to 6-fold for both amaranth grains compared to the control. As the first limiting amino acid was methionine in both grains, its combination with rice containing higher amounts of sulfur-amino acids would improve the amino acid score.

Dietary Fibers and Phenolic Compounds of the Germinated Amaranth Grains

The amounts of soluble, insoluble and total dietary fibers in the A and B grains increased during germination. As for the present experimental method, insoluble dietary fiber contains resistant starch; thus, its amounts with various physiological effects were considered to be increased in the germinated grains compared with the control. Therefore, the germination process of the grains was assumed to improve the functional value (Tosi et al., 2001). The antioxidant activity of the grains increased remarkably during germination. The values of the A and B grains were approximately 6 and 13 times higher, respectively, compared with that of the control. As a result, the germination process for the grains might increase their functionality by changing the dietary fiber levels, as described above (Table 7.2).

Table 7.1. Changes in free amino acid compositions during the germination of amaranth grains

Amino acids	A: Germination time (hr)				B: Germination time (hr)			
	0	24	48	72	0	24	48	72
Arg	11.2	42.5	47.5	57.3	15.6	37.9	44.0	74.3
Met	0.0	3.0	5.2	2.1	4.1	11.0	3.8	4.8
Phe	1.9	3.6	13.1	9.3	3.1	9.1	8.7	10.4
Lys	0.8	2.1	13.2	10.3	1.0	3.6	9.9	24.0
His	5.9	27.4	45.1	39.6	5.0	32.6	51.9	71.3
I-Leu	3.2	5.3	14.4	22.4	5.0	11.9	40.0	62.3
Leu	2.3	2.6	15.1	18.6	3.9	10.3	28.1	58.0
Val	3.6	5.9	20.1	23.7	5.7	12.8	37.5	73.6
Thr	2.9	7.1	17.0	17.1	4.4	16.6	28.6	30.7
Asp+Asn	16.7	115.4	159.8	157.0	15.0	125.7	135.8	262.5
Glu+Gln	40.7	45.9	155.4	213.0	16.1	79.1	525.5	882.7
Tyr	3.7	9.9	25.0	24.5	5.1	23.1	48.8	40.9
Pro	1.1	7.3	4.4	7.2	4.1	16.0	11.5	7.0
Ser	5.1	26.7	30.6	23.5	5.8	40.0	29.6	37.8
Gly	0.5	3.3	3.3	3.9	4.8	4.7	5.8	7.3
Ala	5.4	13.0	18.3	25.2	5.8	21.2	16.5	21.0
Cys	3.5	5.5	3.4	4.7	3.9	4.4	5.0	7.4
Orn	0.0	0.9	1.0	1.2	0.4	0.8	1.4	0.6
GABA	2.2	3.1	6.3	11.7	3.4	9.2	12.5	19.2
Total	111.9	334.7	610.7	681.0	115.6	478.4	1069.8	1843.6

All values are expressed on a dry weight basis (mg/100g).

Ferulic acid is a phenolic compound present in cereal grains and is esterified with the OH-residue of arabinoxylan, thereafter forming a matrix polysaccharide. The content of ferulic acid did not show detectable changes until 24 hr of incubation for both A and B grains, but thereafter, ferulic acid content in the B grain increased distinctly compared to A. The amounts of ferulic acid obtained from the A and B grains were 0.3 mg and 1.0 mg/g grain, respectively, after 72 hr of germination. Based on the relationship between total dietary fiber and ferulic acid content, an increase in the total dietary fiber tended to coincide with an increase in the amount of ferulic acid content. However, there is no such relation in the case of soluble dietary fiber. However, the total polyphenolic content in grain A and B tended to increase during the course of germination. The values of grains A and B were 2.9 times (2.1 mg/g grain) and 6.6 times (3.7 mg/g grain) higher, respectively, as compared with the ungerminated control (Matsuda & Kudoh, 2000). This result suggests that the glycosidic linkage of polyphenol was hydrolyzed during germination under dark conditions, coinciding with the reported results for soybean germination (Lin & Lai, 2006).

Table 7.2. Changes in SDF, IDF and TDF during the germination of amaranth grains

Germination time (hr)	SDF (mg/g)	IDF (mg/g)	TDF (mg/g)	IDF/TDF (%)
A				
0	6.8^{a}	110.6^{a}	117.4^{a}	94.2
24	6.0^{a}	113.0^{a}	119.0^{a}	95.0
48	50.5^{c}	210.1^{bc}	260.6^{c}	80.6
72	47.3^{c}	227.3^{c}	274.6^{c}	82.8
B				
0	11.2^{a}	104.3^{a}	115.5^{a}	90.3
24	9.1^{a}	134.8^{b}	143.9^{b}	93.7
48	9.7^{a}	273.5^{d}	283.2^{c}	96.6
72	27.2^{ab}	312.6^{e}	339.8^{d}	92.0

Abbreviations: SDF, soluble dietary fiber; IDF, insoluble dietary fiber; TDF, total dietary fiber. The values with the same letter are not significantly different according to Scheffe's multiple range test ($P<0.05$). n=3.

VARIATION OF ANTIOXIDANTS DURING GERMINATION

The antioxidant activities of grains A and B were tested during germination (Table 7.3). The values obtained from three antioxidant estimating methods—DPPH (1,1-diphenyl-2-picrylhydrazyl scavenging activity), TEAC (trolox equivalents antioxidant capacity), and SOD (super oxide dismutase)—increased gradually during the germination time course. Grains A and B showed a constant increase in the amount of antioxidant activity. All three antioxidant activities showed a positive correlation with germination time, indicating that these values are derived from polyphenolic compounds in the A and B grains.

As a result, these functional materials were formed during the germination process of the amaranth grains. Therefore, the germination of grain is an easy method for enhancing functional substance levels without the need for a complicated and expensive apparatus.

Table 7.3. Effect of germination on phenolic compounds and antioxidant capacities (DPPH, TEAC, SOD) of amaranth grains

Germination time (hr)	PC (mg/g)	DPPH µmol of VCE/g	TEAC µmol of TE/g	SOD (%)
A				
0	0.72^{a}	0.93^{b}	1.76^{a}	11.0
24	1.12^{b}	2.40^{d}	5.96^{c}	11.0
48	1.76^{c}	4.32^{e}	8.59^{d}	24.7
72	2.08^{d}	5.30^{f}	8.76^{d}	32.3
B				
0	0.56^{a}	0.58^{a}	3.19^{b}	13.8
24	1.27^{b}	1.27^{c}	3.83^{b}	17.6
48	2.92^{e}	6.76^{g}	10.33^{e}	32.9
72	3.70^{f}	7.37^{h}	12.83^{f}	41.0

Abbreviations: PC, phenolic compounds; TEAC, trolox equivalents antioxidant capacity; VCE, vitamin C equivalents.

The values with the same letter are not significantly different according to Scheffe's multiple range test ($P<0.05$). n=3.

EFFECT OF GERMINATION ON THE PROTEIN QUALITIES OF AMARANTH GRAINS

The four fractions of albumin, globulin, prolamin and glutelin were obtained from the grains by a stepwise extraction using distilled water, 5% NaCl-solution, 70% ethanol-solution and 0.02% NaOH-solution. Then, the prepared supernatant was freeze-dried, and the protein contents in each fraction were calculated by the Kjeldahl method.

The protein composition of amaranth changed distinctly after germination (Table 7.4). The amounts of proteins, except for albumin, clearly decreased; this increase was a specific change observed for the germinated amaranth grains. The increase in the amount of albumin could have occurred because many enzyme proteins are composed of albumin (Bewley & Black, 1978; Wanasundara et al., 1999). Although the molecular weight of four types of proteins decreased, prolamin contained only a small amount before germination, and after germination for 48 hr, most of the SDS-PAGE protein bands disappeared (Figure 7.1). This result suggests that most types of proteins degraded into small molecular compounds. Additionally, degradation to a lower molecular weight is generally accompanied by the possibility of reduced allergic response.

Table 7.4. Protein compositions of albumin, globulin, glutelin and prolamin fractions in the amaranth grains

Germination time (hr)	Albumin (%)	Globulin (%)	Glutelin (%)	Prolamin (%)	Residue (%)
A					
0	23.6^{ab}	16.5^{b}	23.6^{f}	0.2	36.1
24	29.3^{b}	16.0^{b}	17.5^{e}	0.1	37.1
48	39.3^{cd}	10.3^{ab}	11.4^{c}	0.1	38.8
72	38.3^{c}	10.6^{ab}	13.3^{d}	0.1	37.8
B					
0	22.1^{a}	14.0^{b}	29.0^{g}	0.2	34.7
24	43.9^{cde}	10.8^{ab}	8.2^{b}	0.1	37.0
48	46.6^{e}	6.1^{a}	6.8^{a}	0.1	40.3
72	45.4^{de}	5.2^{a}	8.9^{b}	0.1	40.4

Each value shows the percentage of total protein.

The values with the same letter are not significantly different according to Scheffe's multiple range test ($P<0.05$). n=3.

SDS-PAGE and Immunoblotting of Germinated Grains

For the analysis of the protein components in the germinated grains, all protein components were analyzed by SDS-PAGE and immunoblotting. To detect allergic reactions to the proteins derived from the grains, immunoblotting analyses were conducted against the sera of patients with wheat-allergic reactions according to Morita et al. (2006). The control sample for allergenic protein was compared with hard-type wheat flour (1CW), and the antigen-antibody reaction was tested. As seen in Figure 7.2, the immunoblotting reaction with proteins from the control grain was very strong and showed three major protein bands. However, with germination, the immune reactions weakened, and these protein bands tended to decrease. In fact, two bands disappeared almost entirely after 48 hr of germination. This tendency was observed for both grains A and B, which suggests a possible lowering of allergenic protein content in the grains during the germination process.

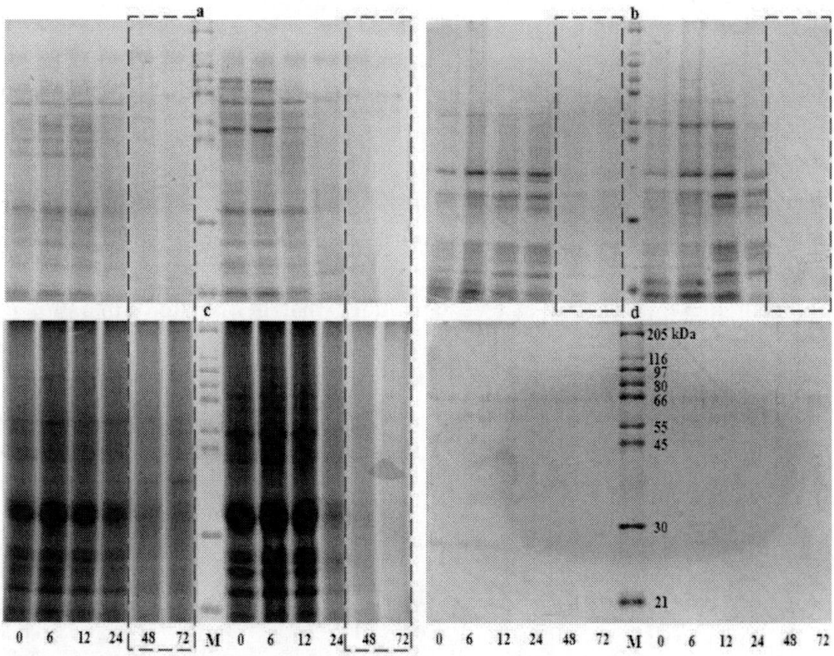

Figure 7.1. SDS-PAGE on albumin (a), globulin (b), glutelin (c) and prolamin (d) fractions during the germination of amaranth grains.
Germination time (hr): 0, 6, 12, 24, 48, 72; M, Marker.

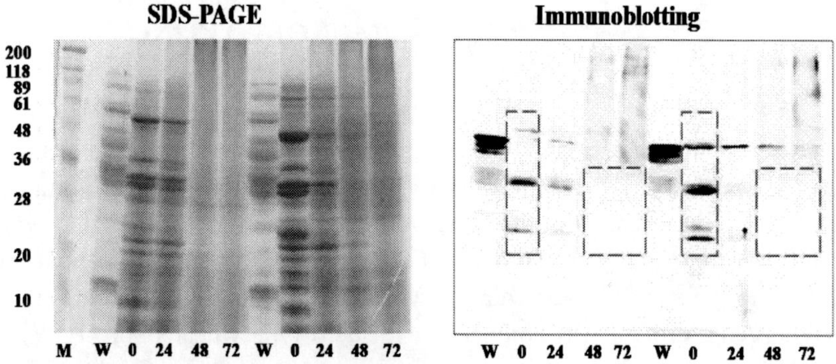

Figure 7.2. SDS-PAGE and immunoblotting analysis of allergenic protein in the amaranth grains during germination.
Germination time (hr): 0, 24, 48, 72; W, Wheat.

Conclusion

During germination, free amino acids, dietary fibers and anti-oxidant activities of the amaranth grains increased with the increase in germination time. The properties of the proteins were also changed by the higher enzyme activity, whereas an increase in the amount of albumin and a decrease in globulin and glutelin were observed. The immunoblotting reactions for the proteins from germinated amaranth against the sera of patients with wheat-allergic reactions were strongly positive compared to the control without germination, but their positive reactions weakened during the germination process.

From these results, it can be concluded that germinated amaranth is a potential substitute for wheat flours. Amaranth flours show promise as a new foodstuff with low-allergenic and high-functional properties compared to wheat, a popular cereal. Moreover, partial substitution of pre-germinated amaranth flour for wheat flour may lower or weaken wheat-allergic reactions and improve the functionality of the final food product.

References

Asao, M., & Watanabe, K. (2010). Functional and bioactive properties of quinoa and amaranth. *Food Science and Technology Research*, 16,

163-168.

Bewley, J. D., & Black, M. (1978). *Physiology and Biochemistry of Seeds.* Springer-Verlag, 306-307.

Kauffman, C. S. (1992). Realizing the potential of grain amaranth. *Food Reviews International.* Marcel Dekker, 8 (1), 5-21.

Lin, P. Y., & Lai, H. M. (2006). Bioactive compounds in legumes and their germinated products. *Journal of Agricultural and Food Chemistry*, 54, 3807–3814.

Matsuda, S., & Kudoh, Y. (2000). Ferulic Acid Contents in mugi miso (Barley-koji miso) and antioxidative activity of mugi koji (Barley-koji). *Food Preservation Science,* 26 (4), 199-203. (*in Japanese*).

Mlakar, S. G., Turinek, M., Jakop, M., Bavec, M., & Bavec, F. (2009). Nutrition value and use of grain amaranth: potential future application in bread making. *Agricultura*, 6, 43-53.

Morita, N., Kang, W. W., Hamauzu, Z., & Sugimoto, Y. (1999). Effect of amaranth flour on some properties of wheat dough and bread. *Journal of Applied Glycosciences*, 46, 23-30.

Morita, N., Maeda, T., Sai, R., Miyake, K., Yoshida, H., Urisu, A., & Adachi, T. (2006). Studies on distribution of protein and allergen in graded flours prepared from whole buckwheat grains. *Food Research International*, 39, 782-790.

National Academy of Sciences (1984). Amaranth: Modern prospects for an ancient crop. National Academy Press, Washington, D. C.

Ogawa, H., Watanabe, K., Mitsunaga, T., & Meguro, T. (2000). Effect of quinoa on blood pressure and lipid metabolism in diet-induced hyperlipidemic spontaneously hypertensive rats (SHR). *Nippon Eiyo Shokuryo Gakkaishi*, 54, 221-227. *(in Japanese*).

Oszvald, M., Tamás, C., Rakszegi, M., Tömösközi, S., Békés, F., & Tamás, L. (2009). Effects of incorporated amaranth albumins on the functional properties of wheat dough, *Journal Science Food Agriculture*, 89, 882–889.

Park, S. H., & Morita, N. (2004). Effect of enzymes on the dough properties and bread quality of wheat flour partly substituted for amaranth flour. *Food Science Technology Research*, 10, 127-131.

Sharpley, C. F. (2009). Malfunction in GABA and glutamate as pathways to depression: A review of the evidence. *Clinical Medicine Insights: Therapeutics*, 1, 1511-1519.

Tosi, E. A., Re, E. D., Lucero, H., & Masciarelli, R. (2001). Dietary fiber obtained from amaranth (*Amaranthus cruentus*) grain by differential milling. *Food Chemistry*, 73, 441-443.

Wanasundara, P. K. J. P. D., Shahidi, F., & Brosnan, M. E. (1999). Changes in flax (Linumusitatissmum) seed nitrogenous compounds during germination, *Food Chemistry*, 65, 289-295.

Section 4: Starch Modification for Functional Food Development

In: Advances in Cereal and Pseudocereal ... ISBN: 978-1-62618-347-6
Editors: N. Morita, P.V. Hung et al.

Chapter 8

STARCH MODIFICATION

***Megumi Miyazaki*[1], *Pham Van Hung*[2] *and Naofumi Morita*[3]**

[1]Research and Development Division, Yamazaki Baking Co., Ltd., Chitose, Sumida-ku, Tokyo, Japan
[2]School of Biotechnology, International University, National University in HoChiMinh City, HoChiMinh City, Vietnam
[3]Department of Food Packaging Technology, Toyo College of Food Technology, Kawanishi, Hyogo, Japan

ABSTRACT

Starches are widely used in food, paper and textile industries; however, starches are unstable during food processing and a number of other industrial treatments. For these reasons, various 'starch modification' techniques have been developed, such as degradation (dextrin), breakage or strengthening of the granules and the improvement of the starch granules, and these starches are termed 'modified starches'. Modified starch granules have been used in many industrial products to improve their quality or to incorporate additional nutritional value. In this chapter, the properties of modified starches, their application in functional foods, such as noodles, confectionaries and bakery products, and the utilization of starches as dietary fiber will be discussed.

1. INTRODUCTION

Starch, the main component of cereal grain and roots, has been widely used in food, paper and textile industries. Starches, however, exhibit a number of shortcomings, and the properties of starches can change easily because of growing or storage conditions. The physicochemical properties of starch are unstable during the food-processing process and other treatments. Therefore, starches are used under controlled heating conditions, or food additives are used to maintain the functional properties of the starch; however, these methods often fall short of expectations. The food industry in particular has attempted to streamline the continuous and large-scale use of starch, and in guaranteeing the uniform quality of the final products, the demands for starch functions have become strict. Furthermore, because the demand for instant, retort-packed and quick-frozen foods has expanded with the changes in lifestyle, the simple selection or blending of starches does not satisfy the demands of the industry.

The properties of starch in food processing, during or after heating, are listed below. These properties represent the shortcomings of native starches.

(1) Insoluble in cold and warm water. The starch does not increase in viscosity following the addition of water.
(2) Swells and gelatinizes when heated. The starch cannot maintain a constant viscosity because the viscosity increases during heating.
(3) Gelatinization is accelerated by starring and pressurization. The viscosity is decreased dramatically by these treatments. Therefore, native starch cannot be used for retort-packed food.
(4) Retrogradation during preservation. Starchy food cannot be stored for long periods, especially at low temperatures or under refrigerated conditions.

For these reasons, various 'starch modification' techniques have been developed, such as degradation (dextrin), breakage or strengthening of the granules and improvement of the granules. Starches can be modified by chemical, physical and enzymatic treatments, which are summarized in Table 8.1 (Ogura, 2004; Thomas & Atwell, 1999a) and are discussed below.

2. Functional Properties of Modified Starches

2.1. Conversion

Starches can be degraded into low molecular weight products by heating, acidity and enzymatic treatments. The conversion process includes pyroconversion (dextrinization), acid hydrolysis, oxidation and enzyme conversion. The conversion process produces starch products containing reduced molecular weight polymers that exhibit reduced viscosity. In general, dextrins do not include oligosaccharides, which exhibit a DP (degree of polymerization) of less than ten.

2.1.1. Pyroconversion (Dry Conditions)

In pyroconversion, the starch products are prepared by dry roasting acidified starch, and the pyroconversion products, which are termed dextrins or more accurately pyrodextrins, exhibit the following properties:

Table 8.1. Classification of modified starches

Type of Modification		Products
Chemical modification	Crosslinking	distarch phosphate, etc.
	Substitution	starch esters: acetylated starch, starch phosphate, octenylsuccinate-treated starch starch ether: hydroxypropylated starch, carboxymethylated starch, cationized starch, etc.
	Conversion	acid-converted starch, oxidized starch, bleached starch pyroconversion (dextrinization): dextrin, British gum, etc.
Physical modification	Pregelatinization	pregelatinized starch
	Heat treatment	heat-moisture treated starch, annealed starch
	Radio treatment	radio treated starch
Enzymatic modification		maltodextrins, cyclodextrin, amylose, etc.

Ogura, 2004; Thomas & Atwell, 1999a.

(1) Extremely low viscosity under gelatinized conditions.
(2) Increased stability of the gelatinized viscosity with the degree of degradation.
(3) Highly concentrated gelatinized liquid exhibiting spinnability.
(4) Gelatinized liquid produces a film, but it is fragile.
(5) Viscosity and adhesiveness are strong.
(6) Highly degraded products are completely soluble in water.

Because of the difficulties associated with the purification of dextrins using water, dry processing is used and this causes dextrins to have a specific smell and taste, which limits the application of dextrins in food processing. Therefore, dextrins are widely used in adhesive and textile industries. Because of their low viscosity, good film-forming ability and high solubility in water, pyrodextrins are used for coating foods and can replace the more costly gums that are used in many of these applications. Specific high viscosity dextrins can be used as fat replacers in bakery and dairy products (Takahashi, 2000a).

2.1.2. Enzymatic or Acid Conversion (Wet Conditions)

Starch can be hydrolyzed into low molecular weight products and ultimately glucose using acids, such as hydrochloric acid or oxalic acid, or enzymatic (alpha-amylase) treatments under conditions that utilize water and heating. The degree of degradation is represented by DE (degree of equivalent), and the theoretic DE of glucose is 100. The products are easy to purify and the cost is low. In the past, roots and tubers were the favored materials used; however, corn starch is now widely used (Takahashi, 2000a).

Enzymatic modification is a common method for the hydrolysis of starch and the final product is dextrin. The enzymatically modified starches (dextrins) are widely used in the food and pharmaceutical industries (Ogura, 2004). The most widely used enzymatic modification of starch is the conversion of the starch to maltodextrins, corn syrups and sugars (e.g., dextrose).

Dextrins are suitable for use as food materials because of their bland taste (although they are somewhat sweet), low viscosity under water-soluble conditions, and the functional properties related to their DE. In general, a change in the DE is accompanied by a change in the functional properties of the dextrins, such as hygroscopicity, solubility, sweetness, viscosity, gelling and water-binding properties. For these reasons, it is possible to design food recipes based on the DE. Dextrins are widely used for solid control (control of water activity), to adjust sweetness, as materials for powders, to improve texture (give thickness) and for the dispersion of pre-mixtures, and because it is a hydrocarbon, dextrin can be used as a type of nutrient (Takahashi, 2000a).

2.2. Physical Modification

2.2.1. Pre-gelatinized Starch

Pre-gelatinized starches (called pregels, precooked starches or instant starches) are "convenience products" that are precooked and dried by the manufacturers and are reconstituted in water to produce a viscous paste (Snyder, 1984). Pregelatinization methods include drum drying, spray cooking, solvent-based processing, extrusion, etc. In most pregels marketed today, the granular integrity has been lost. The instant starches marketed as cold-water swelling types are unique in that they maintain their granular integrity.

To meet the demands of viscosity, moisture retention and adhesiveness, pre-gelatinized starch is used as a food material. Because this starch can form without heating, it is used in foods such as batters for fried food, in cake batter to provide a degree of viscosity, or in boiled fish or meat paste to help the food maintain its form before steaming or boiling. Instant foods or pre-mixtures of bakery cream are other applications of this starch.

2.2.2. Heat-moisture Treated Starch

A less common but interesting physical modification process involves the heat treatment of the starch. Under controlled conditions, granular starch can be heat treated and recovered in its granular form.

There are two types of heat-treatment processes, heat-moisture and annealing treatments, both of which cause the physical modification of the starch without gelatinization, damage to the granular integrity or loss of birefringence (Stute, 1992). For the heat-moisture treatment of starch, the starch is heated to a temperature above its gelatinization point but without sufficient moisture to cause gelatinization. Heat-moisture treated starch improves the consistency and viscosity stability in foods with a pH less than 4.5 (Smalligan et al., 1977).

The annealing of starch involves the heating of slurry of granular starch below its gelatinization point for prolonged periods. After annealing, the granular starch exhibits enhanced viscosity. These starches are considered native starches and are termed "food starch" (Thomas & Atwell, 1999a).

Heat-moisture treated starch granules remain stable under conditions of heating, shearing and high pressure during the cooking processes. Heat-moisture treated starch is used for retort-packed or canned foods, and in pastes for bread, etc. (Suzuki & Sekiya, 1994a, 1994b). The resistant starch (RS) content of high amylose corn starch is increased by the heat-moisture treatment (Azumada, 1997) and acts as a functional material.

2.3. Chemically Modified Starches

The chemical modification of starch was common throughout the last century. Many types of chemically modified starches have been developed and used in the food, paper, and textile industries (Wang, 1997; Endo, 1997; Ito, 1997; Han et al., 2005).

Starch contains abundant hydroxyl groups. Each anhydroglucose unit contains two hydroxyls (at C2 and C3) and a number of anhydroglucose units contain other hydroxyls (at C6). These hydroxyls can potentially react with any chemical that reacts with alcohol hydroxyls, which includes a wide range of compounds, such as acid anhydrides, organic chloro compounds, aldehydes, epoxy, ethylenic compounds, etc. Thus, depending on the purpose of the application, starch can be modified in various ways to improve functionality. Commercially modified starches include hydroxypropylated and/or cross-linked starches and acetylated and/or cross-linked starches, which are widely used in food, textile and paper industries (Ogura, 2004).

2.3.1. Cross-linking of Starch

Cross-linked starches are widely used as thickeners in foods, particularly where a high and stable viscosity is needed. Cross-linking minimizes granule rupture, loss of viscosity and the formation of a stringy paste during cooking. Cross-linking is performed by treating granular starch with multifunctional reagents capable of forming ether or ester linkages with the hydroxyl groups in the starch (Rutenberg & Solarek, 1984; Wurzberg, 1986). When the specific reagent contains two or more moieties capable of reacting with the hydroxyl groups, there is the possibility of reacting at two different hydroxyls, resulting in cross-linking between hydroxyls on the same molecule or on different molecules (Figure 8.1). Cross-linking reinforces the hydrogen bonds in the granule with chemical bonds, which act as a bridge between molecules. As a result, when the cross-linked starch is heated in water, the granule may swell as the hydrogen bonds are weakened; however, the chemically bonded cross-links may provide sufficient granule integrity to keep the swollen granules intact and minimize or prevent a loss of viscosity. Monosodium phosphate (SOP), sodium trimetaphosphate (STMP), sodium tripolyphosphate, epichlorohydrin, phosphoryl chloride (POQ), a mixture of adipic and acetic anhydrides, and a mixture of succinic anhydride and vinyl acetate are reagents that are used to legally cross-link starches (CFR, 1995). The cross-linking of starch improves its stability during cooking, especially under shear and acidic conditions. However, cross-linking also reduces paste clarity and stability during cold storage. Therefore, further modifications, such as hydroxypropylation and acetylation, have been used to decrease the undesirable characteristics of cross-linked starch.

2.3.2. Hydroxypropylation of Starch

Hydroxypropylated starch derivatives formed by the reaction of starch with propylene oxide are used primarily in the food industry. This modification improves the shelf-life, freeze-thaw stability, cold-storage stability, clarity and texture of starch paste. The mechanism for the base-catalyzed reaction of propylene oxide with starch is considered a substitutive nucleophilic bimolecular mechanism and is described in Figure 8.2 (Tuschhoff, 1986). The substitution of propylene oxide is believed to occur primarily at the HO-2-hydroxyl in the starch anhydroglucose unit. A possible reason for this reaction is the high relative acidity of the HO-2 group because of its proximity to the anomeric center. The rate constants for the HO-2, HO-3 and HO-6 hydroxyls are k2:k3:k6 = 33:5:6 (Tuschhoff, 1986).

Hydroxypropyl groups are hydrophilic in nature and when introduced into starch granules, weaken the internal bond structure holding the granule together and prevent the water in the starch paste from separating via syneresis when subjected to freeze-thaw cycling. This reduction in bond strength is reflected in the starch pasting temperature. The higher the level of hydroxypropyl substitution, the lower the pasting temperature until the product exhibits cold-water swelling. Hydroxypropyl groups also prevent retrogradation, resulting in a more fluid paste with improved clarity. This modification can render the desired textural properties in the product (El-Hinnawy et al., 1982; Fleche, 1985; Tuschhoff, 1986).

Cross-linking of hydroxypropylated starch imparts viscosity stability and the desired short-textured property of the paste. Swollen but intact starch granules are desired for most food starch applications to maintain rheological properties. However, for each application, there is an optimum level and balance between hydroxypropyl substitution and cross-linking.

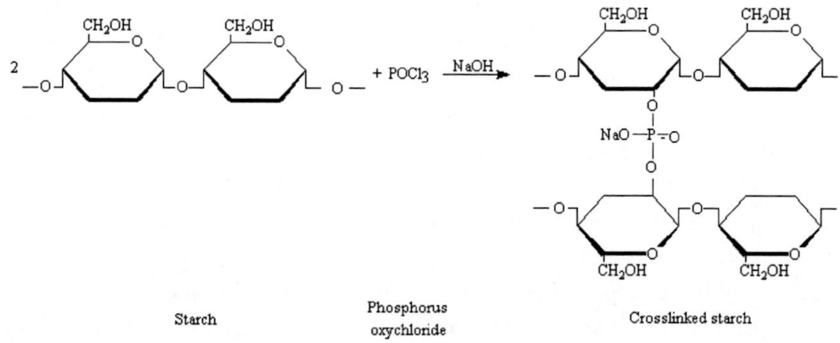

Figure 8.1. Schematic cross-linking of starch (adapted from Miyazaki et al., 2008).

2.3.3. Acetylation of Starch

Acetylated starch has been used for more than a century. For commercial products, starch derivatives of low substitution (degree of substitution (DS) ranging from 0.01 to 0.2, which is approved by the U.S. Food and Drug Administration (FDA) for food) are used to improve binding, thickening, stability and texturing (De Graaf et al., 1998). Acetylated starch with a low DS is generally obtained by the esterification of native starch using acetic anhydride in aqueous medium in the presence of an alkaline catalyst. During acetylation, three free hydroxyl groups, at C2, C3 and C6, in the starch molecule can be substituted with acetyl groups. Therefore, the theoretical maximum DS is three. The expected chemical reactions are illustrated in Figure 8.3 (Mark & Mehltretter, 1972). The reaction shown in Figure 8.3 is the main rate-controlled reaction involving a base catalyst in which starch is acetylated with acetic anhydride and a base catalyst and is an example of nucleophilic substitution at an unsaturated carbon atom of acetic anhydride. The three free OH groups exhibit different reactivities. The primary OH at C6 is more reactive and because of steric hindrance, is acetylated more readily than the secondary OH groups at C2 and C3. The primary OH located at the exterior surface of the starch molecules reacts readily with the acetic groups, whereas the two secondary OH groups located within the interior surface of the starch form hydrogen bonds with the OH groups on the neighboring glucose unit. Regarding the two secondary OH groups, the OH at C2 is more reactive than the OH at C3, mainly because C2 is closer to the hemi-acetal and more acidic than C3. When acetyl groups are introduced into starches, the tendency of solutions of these starches to deteriorate in clarity and texture and to undergo syneresis when held at low temperatures is greatly reduced. These substituent groups stabilize starches by interrupting the linearity of the amylose and segments of the amylopectin branches. The groups interfere sterically with the intermolecular alignment (Harper, 1981). Therefore, acetylated starch typically exhibits physicochemical characteristics, such as low gelatinization temperature, high solubility, good cooking properties and storage stability and is widely used in the food industry to overcome undesirable changes in the product texture and appearance caused by retrogradation during processing and storage.

StOH + NaOH → StO-Na + H_2O
Starch Alkali Sodium starchate Water

HOH
StO-Na + CH2 - CHCH3 → StOCH2CHCH3 + NaOH
O \ /
Propylene Oxide Hydroxypropylated starch

Figure 8.2. The mechanism for the base-catalyzed reaction of propylene oxide with starch (adapted from Miyazaki et al., 2008).

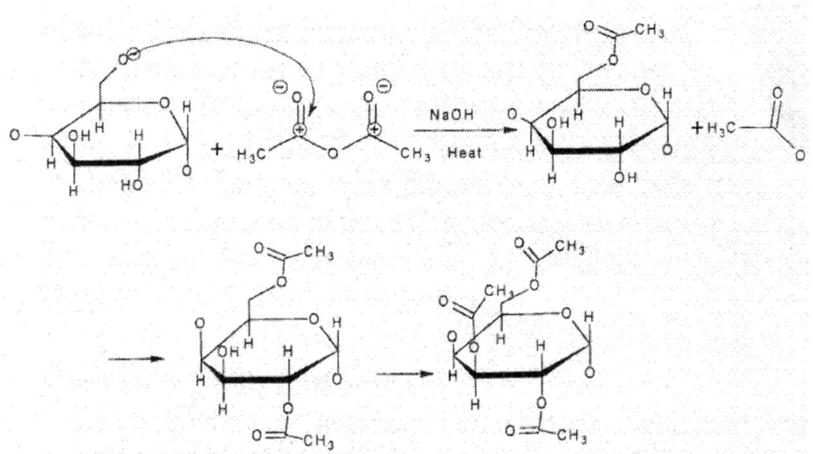

Figure 8.3. Chemical reactions during starch acetylation (adapted from Miyazaki et al., 2008).

3. APPLICATION OF MODIFIED STARCHES IN FUNCTIONAL FOODS

Starches, particularly modified starches, have been widely used as essential food materials (Table 8.2) (Davies, 1995). Starch contributes to texture, viscosity, gel formation, adhesion, binding, moisture retention, film formation and product homogeneity. In addition, certain modified starches are used increasingly as fat substitutes in low- and no-fat products.

Table 8.2. Starch usage by application sector

Application	Binders	Viscosifiers	Film Formers	Texturizers
Soup, sauces, and gravies	…	X, XS, PX, PXS	…	X, XS, PX, PXS
Bakery	PN	X, P, PX, PXS	D,M	P, X, PX, PXS, M
Dairy	N, A, M	X, XS, P, PX, PXS	…	X, XS, PXS, A, PX, O, PO, M
Confectionary	PO, O	…	O, PO, A	…
Snacks	N, P, PN, PO, D	…	…	…
Batters and coatings	X, PX, O	P, PX	D	O, PO, D, M
Meat products	N, X, XS, P	…	XS	XS

Source: Davies, 1995.

Abbreviations: N=native, X=crosslinked, P=pregelatinized, S=substituted, O=oxidized, A=acid hydrolyzed, D=dextrin, and M=maltodextrin.

Commercial starches (Table 8.3) are obtained from cereals, such as corn, wheat, various types of rice and from tubers or roots, such as potato and cassava (tapioca starch). Starches from different sources vary in taste and viscosity properties. For example, native potato and tapioca starches exhibit weak intermolecular bonding and gelatinize easily to produce a high-viscosity paste that thins rapidly with moderate shear. Potato starch produces a clear viscous, almost bland paste, which is used in products such as extruded cereals, dry soup and cake mixes. Tapioca starch produces a clear, cohesive, paste that gels slowly over time. Native corn, rice and wheat starches form opaque, gelled pastes that have a slight flavor of cereal. High-amylose corn starches produce a clear cohesive paste (Whistler & BeMiller, 1997).

Because the paste and gels produced by native starches are often cohesive (gummy) or rubbery, the functional properties of these starches are improved by modification. Different types of modifications produce starches that withstand better the heat, shear and acid associated with various foods and food-processing conditions. Modifications can affect starch solubility, viscosity, freeze-thaw stability, paste clarity and gloss, gel formation, film formation and cohesiveness.

Numerous factors must be considered when choosing a starch for use in a particular food. The desired properties of the food (e.g., texture, mouth-feel and viscosity), and the processing method must be examined. An up-front strategy in which the various requirements of the food product are reviewed prior to the selection of a starch saves time, frustration, and money in the product-development process (Thomas & Atwell, 1999c).

Recently, consumers demand not only good tastes, convenience, and rationalization but also improvement in health. A new trend is emerging in which starch is used as a functional material following its modification (Table 8.4) (Takahashi, 2000c). The functions of modified starches are as follows:

(1) Nutrition: Starch is a hydrocarbon with an energy content of 4 kcal/g. Digestibility is improved by gelatinization or pyroconversion. When dieting, people control their energy intake, and indigestible dextrins or dextrins that substitute for oil and fat are useful for this purpose.
(2) Taste: Taste is a main function of starch, and appearance (gloss), flavor and texture contribute to this property; consumers tend to demand a light taste and soft texture.
(3) Physiology: This is a new function for starch. Starch products, such as oligosaccharides, indigestible dextrins, and sugar alcohol regulate the functions of the intestines and the blood glucose levels and reduce cholesterol. These products are included in foods with specific health uses.
(4) Food processing: The powdering of seasonings or fats and emulsified foods are popular, and modified starches or dextrins are used in these foods.

Table 8.3. Approximate properties of common food starches

Property	Wheat	Dent corn	Waxy corn	High-amylose corn	Rice	Potato	Tapioca
Source	cereal	cereal	cereal	cereal	cereal	tuber	root
Size and Shape							
Diameter (μm)	1-45	5-30	5-30	5-30	1-3	5-100	4-35
Shape	round, lenticular	polygonal, round	polygonal, round	polygonal, round, irregular	polygonal, specific compound granules	oval, spherical	oval, truncate, "kettle drum"
Granule structure							
Amylose (%)	20	25	<1	55-70 (higher)	19	25	19
Amylopectin (%)	80	75	>99	45-30 (lower)	81	75	81
X-ray diffraction pattern	A	A	A	A	A	B	C (similar A)
Gelatinization temperature (°C)							
By microscopic examination	52-63	62-72	63-72	-	61-78	56-66	59-70
To, Tp, and Tr by DTA	57, 64, 82	63, 70, 80	64, 71, 85	-	69, 76, 88	59, 65, 79	67, 73, 85
at peak viscosity by Amylograph	-	88	73	-	93	74	73

Thomas & Atwell, 1999b; Takahashi, 2000b.
-: data not shown.

Grain can be broken down into four fractions, albumin, globulin, prolamin and glutelin, and these fractions are obtained from grains by a stepwise extraction method that uses distilled water, a 5% NaCl solution, a 70% ethanol and 0.02% NaOH solution. Using this method, the supernatant can be freeze-dried and the protein content of each fraction can be calculated using the Kjeldahl method.

Noodles are made from wheat flour or buckwheat flour, and starch is not generally used as a material for noodles, with several exceptions. For example, harusame (gelatin noodles and Japanese vermicelli) is made from green bean, sweet potato, potato or edible canna starch. In Korea, cold noodles are made from wheat flour and potato starch, and rice vermicelli is made from rice powder and starch.

A trial to use starch for producing noodles was initiated in the 1970s, and the first application was in instant noodles to which potato starch was added (Hishikawa, 1996). Next, tapioca starch was used to improve texture. With improvements in the development of distribution systems, sterilization techniques and the increased demands for consumer convenience, frozen or boiled-instant noodles were developed. Modified starch is an essential material for these types of noodles, and it has become one of the main uses of modified starch. The reasons for using starch in noodles are to improve the texture and cooking process, to delay retrogradation, or to delay the onset of sogginess after cooking, as described below (Inada, 2004).

(1) Improve texture: The addition of 5 or 10% starch to wheat improves noodle texture. The surface of the noodle becomes smooth by the addition of starch, which makes it easier to eat. Waxy corn starch produces the softest noodles, potato starch produces elastic and hard noodles, and tapioca starch produces noodles that are elastic but soft. Corn starch produces a noodle with a hard and dry texture; therefore, it is rarely used. These textural tendencies can be produced using modified starches derived from the same materials.
(2) Improve the cooking process: Consumers demand a shorter boiling time not only for instant noodles but also for dry or raw-type noodles. The addition of modified starches that exhibit a low gelatinization temperature is effective. Treatment with higher ether levels or ester modification makes the gelatinization temperature lower.
(3) Delay retrogradation: Freezing is effective way to prevent moisture migration; however, noodles still retrograde during preservation. Starch ester or ether can improve preservation. Long-life noodles (raw-type instant noodles) can be preserved for a long time; however, retrogradation and moisture uniformity can occur. To control this phenomenon, the inclusion of 20 to 40% starch ester or ether is effective.
(4) Delay texture changes: Noodles become soggy after boiling or soften during cooking. To delay these phenomena, the use of cross-linked starch is effective. Cross-linked starch produces a hard texture and although it is not preferable to use cross-linked starch alone, it is effective when combined with other starches. The total amount of starch need for this effect is between 5 and 30%.

3.2. Confectionaries

3.2.1. Gum Candies

Gum candies (e.g., gumdrops and jellies) are soft with a chewy consistency. The two types of starch most commonly used in the confection industry are high-amylose starch and thin-boiling (acid-thinned) starch. The latter is acid-hydrolyzed starch, which is used to reduce the hot viscosity of a paste so that higher concentrations of starch can be dispersed without excessive thickening. These starches can be used alone or combined at various ratios depending on the texture desired. The total percent starch content is approximately 9-15% by weight (Thomas & Atwell, 1999d).

High-amylose starch, when properly cooked, has high gel strength and sets up rather quickly. This rapid set improves productivity; however, high amylose starch has one limitation, if it is cooked at too high a temperature, tailing during deposition and rapid gelling, which can cause problems in the holding tank, can occur. For these reasons, high-amylose starches are often acid-thinned to reduce viscosity or are used in combination with other thin-boiling starches.

3.2.2. Pan Coating

During pan coating, candies such as jelly beans are rotated in a revolving pan and coated with a concentrated sugar solution containing a film-forming agent. High-amylose, converted starches are often used in this coating process because of their moisture-absorbing and film-forming abilities (Thomas & Atwell, 1999d).

3.2.3. Hard Candies

Larger amounts of starch (typically 20-30% on a dry solids basis) are used in the manufacture of hard candies. Because of the larger amount of used, the starches must be lower in viscosity than in soft, jelly-type candies. Very low viscosity, thin-boiled starches and dextrins (DE 15-20) can be used to provide a pliable toughness to the candy pieces rather than the cuttable gel structure required for soft candies (Thomas & Atwell, 1999d).

3.2.4. Nougats

Nougats are made from a combination of sugars, syrups, and whipping agents. There are two basic types of textures (short and chewy), which are controlled by using different amounts of non-crystalline sugars and the addition of a specialty starch. The starch differs depending on the property desired. When chewiness is desired, an unmodified waxy starch (usage level approximately 2%), or native tapioca starch are among the types of starches that can be used. When stiffness, body, and textural stability are desired, a modified specialty starch (approximately 3%) derived from waxy starch can be used (Thomas & Atwell, 1999d).

3.3. Bakery Products

3.3.1. Bread

Bread is made mainly from wheat flour. When water containing sodium chloride is added and is mixed with the flour, the water-soluble proteins together with the water-insoluble proteins hydrate and form gluten; starch and other dough components are embedded (Beccari, 1745). Thus, the gluten is essentially the skeleton or the framework of the resulting dough within which gas retention is achieved, which produces light and leavened products. Because gluten protein plays an important role in breadmaking, cereal chemists have paid many attention to insoluble wheat proteins, namely gliadin and glutenin (Uthayakumaran et al., 2000a, 2000b, 2000c; Cup, 2000; Wesley, 2001).

Wheat protein makes up approximately 12% (anhydrous basis) of the endosperm of the grain, whereas starch is approximately 80% (anhydrous basis) of the grain. Most of the starch is embedded in the gluten and becomes the skeleton of bread, whereas a number of carbohydrates are hydrolyzed by yeast or by native enzymes and are used as a substrate for the fermentation process, yielding mainly ethanol and carbon dioxide (Kulp, 1973a, 1973b). In addition to the role of gluten, the role of starch in baked products is relatively important.

The decrease in bread quality is due in part to the decrease in the flavor impact or a drier texture during storage, which is called "staling" or bread staling (Stauffer, 2000). Factors affecting bread staling have been investigated extensively, and the rearrangement of the starch fraction was found to be a significant cause (Hug-Iten et al., 2003). Therefore, a number of additives, such as fat, emulsifiers, sugars and α-amylase, which can retard starch retrogradation and prolong the shelf life of bread have been widely used in the bread-baking industry.

Recently, modified starches have been developed extensively and are widely used in the production of starchy foods, such as noodles, cakes, snacks and flour paste, to retard retrogradation, improve food quality, or provide additional nutritional value (Wang, 1997; Endo, 1997; Wada, 1997; Ito, 1997; Han et al., 2005). Similar to noodles, cakes, snacks and flour paste, bread is a starchy food, and modified starch is used to improve the shelf life and functional properties of bread. Recent studies have demonstrated that a number of chemically modified starches, such as cross-linked hydroxypropylated potato starch (Takasaki & Mineki., 2001), hydroxypropylated tapioca starch, cross-linked (Miyazaki, et al., 2005a, 2005b) and a number of cross-linked waxy starches (Hung et al., 2004), are useful ingredients for delaying the staling of bread.

3.3.2. Cookies

Cookie formulas vary depending on the desired final product. Cookie ingredients typically include flour, water, eggs, sugars, fats and leavening agents. The starch in cookies is derived from the flour, unless starches are added for specific functions. The cohesiveness of cookie dough and the framework of the final product are derived from the starch. In a typical cookie formula, the native starch remains almost entirely ungelatinized after baking and occurs as intact granules. The starches do not gelatinize because of the presence of sugars, which compete with the starches for the water needed for gelatinization.

Often, starches are added to formulations as tenderizers to create a soft, chewy texture or for crispiness. In addition, modified waxy starches or precooked starches are often used to increase the water-holding properties of the dough. These starches aid the formation of a tender structure and reduce the stickiness of the dough, which can aid in the molding and cutting process. Finally, modified starches that control swelling during heating are used to generate crispiness in the final product, such as in crackers and pretzels (Thomas & Atwell, 1999e).

3.3.3. Cakes

Cake batters are emulsions of starches, proteins, sugars, water and fats. The basic formulations for cake batter contain flour, sugar, eggs, shortenings, leavening agents and a liquid such as water or milk. Starches in these formulas serve several functions. Following heating, the granular starches in the formula swell (approximately 80°C), absorbing water, which increases the viscosity and therefore the stability of the batter emulsion. Cake-type products contain enough water to gelatinize the starches, although a portion of ungelatinized granules may remain. The gelatinized starches form a matrix in which air bubbles are entrapped. This matrix is the major structural component of the cake-cell structure. The timing of the gelatinization of the starches is important to ensure the correct final cake structure and crumb development. The timing of the gelatinization can be controlled by changing the level of sucrose or other sugars, such as dextrose, in the formula or by adding unmodified or modified starches, depending on the formulation (Thomas & Atwell, 1999e).

Typically, cakes are preferable when they contain fine grains, a large volume, smooth texture, and slow staling, and these properties ensure less shrinkage after baking. Starches are added to batter formulations to improve the quality of the final products. Starches are often added as tenderizers (at approximately 1-5%) to soften the crumb and as thickeners to increase the viscosity of the batter, which improves the quality of the air cell size and increases the quality of the final product (Thomas & Atwell, 1999e). A number of methods are used to reduce shrinkage, such as the physical method, which entails a shock that standardizes the size of each air cell, or the addition of emulsifiers, heat-treated or bleached flour, and pregelatinized starch. Modified starches that control swelling, such as cross-linked tapioca (at approximately 20-30%) are used as novel ingredients to reduce shrinkage. The use of cross-linked starches in dry cake mix applications also helps to provide shear resistance, which improves the tolerance of the batter to over mixing by the consumer. The addition of starches also aids in moisture retention in the final product (Takahashi, 2000c).

Modified starches can be used as fat mimetics or to replace fats and are often used in reduced-fat cake applications.

3.3.4. Snacks

There are many types of snacks. Regarding the ingredients of snacks, wheat flour is used for cookies and pretzels, corn and potato flours are used in chips and rice flour is used in rice crackers. An important characteristic of snacks is their crispy texture, and starches play an important role in these products. The degree of gelatinization is different for each of these products; the starch granules remain unchanged in cookies, whereas the starches swell and extend in chips. This effect is influenced by the water in the product and by the type of starches used (Takahashi, 2000c). Cookies were discussed in a previous section (3.3.2); therefore, this section will focus on chips, especially the extruded products described below.

Starches consist of amylose and amylopectin. The amylose is responsible mainly for the strength of the mixing process and the crispiness of the extruded products. The amylose in formulas makes the dough resistant to shear degradation during extrusion and can help to improve cutting and shape retention during drying, baking or flying. Gelatinized amylose forms a tight polymer network and contributes the main structure of the products. Amylopectin forms a web-like network through gelatinization. The network increases the product viscosity and forms a structural framework for the expansion process. The setting of the framework aids in the generation of a crisp final product. Because of polymer branching, amylopectin is susceptible to shear degradation during extrusion. The resulting dextrins and short-chain polymers cause an increase in stickiness and difficulties in the cutting process (Thomas & Atwell, 1999e).

Table 8.4. Function of food and starch or its products

Function		Effects	Starch or degraded starch
Nutrition	Control shortage	Aid digestion absorption	dextrins
	Control excess	Low calories	indigestible dextrins, oligosaccharides
Tastes	Color	Gloss	dextrins
	Flavor	Encapsulated	porous dextrins, cyclodextrin
	Taste	Savory flavor	starch degraded
	Texture	Viscosity	modified starch
		Gelling	modified starch
		Puffed	modified starch
Physiology	Biorhythm	Intestinal regulation	indigestible dextrins, oligosaccharides
	Prevention of and recovery from disease; adiposeness, diabetes, cancer, cavity	Reduce insulin secretion	indigestible dextrins
		Decrease serum cholesterol	indigestible dextrins
		Decline in high blood pressure	indigestible dextrins
		Prevent a cavity	oligosaccharides, sugar alcohol
Food process		Powdering	dextrins
		Emulsifying	modified starch
		Film forming	modified starch
Preservation		Water activity	starch degraded
		Water separation	modified starch
		Retrogradation	modified starch
		Stability for base or acid	modified starch

Takahashi, 2000c.

Native starches are generally not resistant to high temperatures and shear, and the breakdown of starch results in a sticky dough. However, light or moderately cross-linked starches can help alleviate these problems.

3.4. Frozen Foods

Frozen foods are preserved and distributed at temperatures below -18°C, and degradation in the quality of frozen foods is considered small; however, there are a number of problems associated with frozen foods.

Whereas the protein in foods is denatured by heating, freezing has little effect on the food quality, with the exception of tofu and eggs. However, freezing retrogrades gelatinized starch. To delay the deterioration of the gelatinized starch, the addition of emulsifiers, sugars, waxy corn starch or waxy rice is effective. Modified starches, such as starches cross-linked with or without esters or corn starch modified using ether are also effective because they have a minimal effect on the flavor of the foods (Takahashi, 2000c).

3.5. Retorted Products

A retort is essentially a commercial-scale pressure cooker that is used in canning operations to process food thermally at high temperatures (e.g., 120°C) and to increase pressures for various lengths of time. Because of these severe treatments, retorted products are liable to change color, texture, or viscosity, especially in low-acid foods. To prevent these problems, heat-stable starches, such as cross-linked or heat-moisture treated corn starches are used. Because heat-moisture treated starch is subjected to heat and water under vapor pressure and chemical reagents are not used, this modified starch is considered a natural ingredient.

There are two types of starch used for retorted products, the so-called "fill viscosity" starch, and the "thin-thick" starch. The "fill viscosity" starch gelatinizes easily prior to retorting to give viscosity during the initial mixing, cooking and filling stages of the canning operation. The thin-thick" starch does not develop viscosity until the system has reached the high temperatures associated with the retorted process. The appropriate starch must be selected for every retorted product (Thomas & Atwell, 1999f).

Table 8.5. Classification or resistant starch

Category	Definition
Type I	Physically inaccessible starch that is locked in plant material such as milled grains, seeds, and legumes
Type II	Native granular starch found in uncooked food such as bananas
Type III	Indigestible starch that forms after heat and moisture treatment; may be present in foods such as cooked potatoes
Type IV	Resistant starch that is produced by chemical or thermal modification

Thomas & Atwell, 1999g.

3.6. Use of Starch as Dietary Fiber

Until the early 1980s, starch was thought to be digested completely in the small intestine; however, in 1982, Englyst et al. discovered “resistant starch (RS)”, defined as dietary starch that is not digested in the small intestine (Englyst et al., 1982).

There are four types of RS (Table 8.5) (Thomas & Atwell, 1999 g). In Type I starches, the cell walls or other barriers protect the starch from the action of the digestive enzymes and acidic conditions of the digestive tract. Highly crystallized native starches, such as those found in raw bananas or potatoes, cannot be reduced into degradation products that can be absorbed through the small intestine and are considered Type II starches. A Type III resistant starch is created if the starch in a food system is gelatinized, cooled and allowed to retrograde. Finally, it is possible to generate a starch that is resistant to digestion by modifying and altering a native starch (Type IV starches). The most popular RS is high-amylose corn starch, which contains approximately 30% dietary fiber, and the heat-moisture treatment of high-amylose corn starch increases the RS by 60 or 70% (Osumi, 2004). Heat moisture-treated starches are used in bakery products and noodles.

Indigestible dextrin is a water-soluble dietary fiber derived from dextrins (pyroconversion). This starch has a low viscosity, slight sweetness and an energy content of only 1.0 kcal/g. Dextrin exhibits several physiological effects similar to many other dietary fibers, namely, it is not digested in the small intestine but is utilized by bifidobacteria in the colon to produce short-chain fatty acids and to increase feces production. Dextrin reduces the amount of sugar that the body can absorb, resulting in the moderation of rapid increases in blood glucose levels. It has been reported that dextrin can decrease the natural fat and cholesterols in the blood serum and prevents obesity when it is consumed regularly (Osumi, 2004). Because of the properties described above, indigestible dextrin is used in foods consumed for specified health reasons (Takahashi, 2000c).

CONCLUSION

The food, paper and textile industries have increased the development of modified starches. To produce modified starches, a number of the shortcomings of native starches, such as insolubility in water, unstable viscosity during heating, high shear or pressure and degradation, has been improved, therefore modified starches are widely used in place of native starches in food industries to improve the quality or to control the food processing procedure. The most advantageous properties of starches used in the production of starchy foods is the texture they provide, and starches are used in confectionaries, bakery products, udon noodles and retorted or frozen foods to provide a specific texture or to delay degradation. Starch products, such as oligosaccharides, indigestible dextrins, resistant starches and sugar alcohol, regulate the functions of the intestines and blood glucose levels and reduce cholesterol. Starch products are used in a number of foods for specified health reasons and it is expected that in the future, the use of these products will increase.

REFERENCES

Azumada, K., & Nakakuki T. (1997). Properties and applications for heat-moisture treated starches. *Food Chemicals*, 10, 35-40 (in Japanese).

Beccari, A. (1745). *De frumento. Instituto atque academia commentarii* 2, Part 1, p. 122.

CFR (Code of Federal Regulations) (1995). Food starch modified. In: *Food Additives Permitted in Food for Human Consumption* (title 21, chapter 1, part 172, section 172.892). Washington: US Government Printing Office.

Cup, B., Boutrot, F., Redl, A., & Lullien-Pellerin, V. (2000). Study of the temperature effect on the formation of wheat gluten network: Influence on mechanical properties and protein solubility. *Journal of Agriculture and Food Chemistry,* 48, 2954-2959.

Davies, L. (1995). Starch-composition, modification, applications and nutritional value in foodstuffs, In: D. J. Thomas and W. A. Atwell (Eds), Starches: Practica Guides for the Food Industry, Eagan Press, St. Paul, pp.44-52

De Graaf, R. A., Broekroelofs, A., & Janssen, L. B. M. (1998). The acetylation of starch by reaction extrusion. *Starch/Staerke*, 50, 198-205.

El-Hinnawy, S. I., El-Saied, H. M., Fahmy, A., El-Shirbeeny, A. E., & El-Sahy, K. M. (1982). Viscosity and gelatinization characteristics of hydroxyethyl starch. *Starch/Staerke*, 34, 112-114.

Endo, Y. (1997). Utilization of chemically modified starches to snack food. *Food Chemicals*, 13, 41-47 (in Japanese).

Englyst, H., Wiggins H.S., & Cummings J. H. (1982). Determination of the non-starch polysaccharides in plant foods by gas-liquid chromatography of constituent sugars as alditol acetates. Analyst, 107, 307-318.

Fleche, G. (1985). Chemical modification and degradation of starch. In: G. M. Van Beynumy & J. A. Roel, (Eds), *Starch Conversion Technology*, (pp.73–87). New York, USA: Marcel Dekker Inc.

Han, J.-A., Lee, B.-H., Lim, W. J., & Lim, S.-T. (2005). Utilization of hydroxypropylated waxy rice and corn starches in Korean waxy rice cake to retard retrogradation. *Cereal Chemistry*, 82, 88-92.

Harper, J. M. (1981). Food Extrusion. In: J. M. Harper, (Ed), *Extrusion of Foods*, Vol.1 (pp.1-3). Boca Raton: CRC Press.

Hishikawa, Y. (1996). Recent developments and applications of modified starches as food materials. *Japan Food Science*, 35, 23-28 (in Japanese).

Hug-Iten, S., Escher, F., & Conde-Petit, B. (2003). Staling of bread: Role of amylose and amylopectin and influence of starch-degrading enzymes. *Cereal Chemistry*, 80, 654-661.

Hung, P. V., & Morita, N. (2004). Dough properties and bread quality of flours supplemented with cross-linked cornstarches. *Food Research International*, 37, 461-467.

Inada, K. (2004). Modified starch and utilization. In F. Hidetsugu, T. Komaki, S. Hizukuri, & K. Kainuma (Eds), *Encyclopedia on starch-science*, (pp. 504-505). Tokyo: Asakura-shoten (in Japanese).

Ito, T. (1997). Application of starch that has freeze-thaw stability to processed food. *Food Chemicals*, 13, 58-62 (in Japanese).

Kulp, K. (1973a). Properties of starch granules derived from flours and parent wheats. *Baker's Digest*, 47 (1), 55-59.

Kulp, K. (1973b). The physicochemical properties of wheat starch as related to bread. *Baker's Digest*, 47 (5), 34-38.

Mark, A. M., & Mehltretter, C. L. (1972). Facile preparation of starch triacetates. *Starch/Staerke*, 24, 73-76.

Miyazaki, M., Maeda, T., & Morita N. (2005a). Gelatinization properties and bread quality of flours substituted with hydroxypropylated, acetylated and phosphorylated cross-linked tapioca starches. *Journal of Applied Glycoscience*, 52, 345-350.

Miyazaki, M., Maeda, T., & Morita N. (2005b). Starch retrogradation and firming of bread substituted with hydroxypropylated, acetylated and phosphorylated cross-linked tapioca starches for wheat flour. *Cereal Chemistry*, 82, 639-644.

Miyazaki, M., Hung, P. V., & Morita, N. (2008). Recent advances in application of modified starches for breadmaking. *Trends in Food Science and Technology*, 17(11), 591-599.

Ogura, T. (2004). Modified starch and utilization. In: F. Hidetsugu, T. Komaki, S. Hizukuri, & K. Kainuma (Eds.), *Encyclopedia on starch-science*, (pp. 393-427). Tokyo: Asakura-shoten (in Japanese).

Osumi, K. (2004). Utilization as fibers. In: F. Hidetsugu, T. Komaki, S. Hizukuri, & K. Kainuma (Eds.), *Encyclopedia on starch-science*, (pp. 511). Tokyo: Asakura-shoten (in Japanese).

Rutenberg, M. W. & Solarek, D. (1984). Starch derivatives: production and uses. In R. L. Whistler, J. N. BeMiller, & E. F. Paschall (Eds.), *Starch: Chemistry and Technology* (pp. 312-388). New York: Academic Press.

Smalligan, W. J., Kelly, V. J., & Enad, E. J. (1977). Preparation of a stabilized precooked baby food formulation thickened with modified tapioca starch. U.S. patent 4, 013, 799.

Snyder, E. M. (1984). Industrial microscopy of starches. In R.L. Whistler, J. N. BeMiller & E. F. Paschall (Eds.), Starch: Chemistry and Technology, 2nd ed (p.670). Orland, Florida: Academic Press.

Stauffer, C.E. (2000). Emulsifiers as antistaling agents. *Cereal Foods World*, 45, 106-110.

Stute, R. (1992). Hydrothermal modification of starches: The difference between annealing and heat-moisture treatment. *Starch/Staerke*, 44, 205-214.

Suzuki, A., & Sekiya, S. (1994a). Application of heat-moisture treated starch, 'derica star' to food materials. I: The application to pouch-packed and canned foods. *Food Chemicals*, 10 (2), 67-73. (in Japanese).

Suzuki, A., & Sekiya, S. (1994b). Application of heat-moisture treated starch, 'derica star' to food materials. II: The application to fillings. *Food Chemicals*, 10 (3), 93-98. (in Japanese).

Takahashi, R. (2000a). Starch modification. In R. Takahashi, (Ed.), *Knowledge of starch* (pp. 92-124). Tokyo: Saiwai Shobo (in Japanese).

Takahashi, R. (2000b). Science of starch. In R. Takahashi, (Ed.), *Knowledge of starch* (p. 63). Tokyo: Saiwai Shobo (in Japanese).

Takahashi, R. (2000c). Starch utilization. In R. Takahashi, (Ed.), *Knowledge of starch* (pp. 125-218). Tokyo: Saiwai Shobo (in Japanese).

Takasaki, S., & Mineki, M. (2001). Effects of potato starch and modified potato starch on breadmaking properties. *Journal of Cookery Science of Japan*, 34, 53-61 (in Japanese).

Thomas, D. J., & Atwell, W. A. (1999a). Starch modifications. In D. J. Thomas, & W. A. Atwell (Eds.), *Starches* (pp. 31-48). St. Paul, Minnesota: American Association of Cereal Chemists.

Thomas, D. J., & Atwell, W. A. (1999b). Starch structure. In D. J. Thomas, & W. A. Atwell (Eds.), *Starches* (pp. 1-11). St. Paul, Minnesota: American Association of Cereal Chemists.

Thomas, D. J., & Atwell, W. A. (1999c). Matching starches to applications. In D. J. Thomas, & W. A. Atwell (Eds.), *Starches* (pp. 49-66). St. Paul, Minnesota: American Association of Cereal Chemists.

Thomas, D. J., & Atwell, W. A. (1999d). Confections and dairy products. In D. J. Thomas, & W. A. Atwell (Eds.), *Starches* (pp. 57-64). St. Paul, Minnesota: American Association of Cereal Chemists.

Thomas, D. J., & Atwell, W. A. (1999e). Grain-based procucts. In D. J. Thomas, & W. A. Atwell (Eds.), *Starches* (pp. 65-74). St. Paul, Minnesota: American Association of Cereal Chemists.

Thomas, D. J., & Atwell, W. A. (1999f). Sauces, dressings, and other applications. In D. J. Thomas, & W. A. Atwell (Eds.), *Starches* (pp. 75-82). St. Paul, Minnesota: American Association of Cereal Chemists.

Thomas, D. J., & Atwell, W. A. (1999g). Special topics. In D. J. Thomas, & W. A. Atwell (Eds.), *Starches* (pp. 83-87). St. Paul, Minnesota: American Association of Cereal Chemists.

Tuschhoff, J. V. (1986). Hydroxypropylated starches. In O. B. Wurzberg (Ed.), *Modified Starches: Properties and Uses* (pp. 89-96). Boca Raton: CRC Press.

Uthayakumaran, S., Stoddard., F.L., Gras, P.W., & Bekes, F. (2000a). Optimized methods for incorporatinig glutenin subunits into wheat dough for extension and baking studies. *Cereal Chemistry*, 77, 731-736.

Uthayakumaran, S., Stoddard., F.L., Gras, P.W., & Bekes, F. (2000b). Effects on incorporated glutenins on functional properties of wheat dough. *Cereal Chemistry*, 77, 737-743.

Uthayakumaran, S., Newberry, M., Keentok, M., Stoddard., F.L., & Bekes, F. (2000c). Basic rheology of bread dough with modified protein content and glutenin-to gliadin ratio. *Cereal Chemistry*, 77, 744-749.

Wada, K. (1997). Improvement of bread and cake quality with starches. *Food Chemicals*, 13, 30-34 (in Japanese).

Wang, S. W. (1997). Starches and starch derivatives in expanded snacks. *Cereal Foods World*, 42, 743-750.

Wesley, A. S., Lukow, O.M., Mckenzie, R.I.H., Ames, N., & Brown, D. (2001). Effect of multiple substitution of glutenin and gliadin proteins on flour quality of Canada prairie spring wheat. *Cereal Chemistry*, 78, 69-73.

Whistler, R. L., & BeMiller, J.N. (1997). Carbohydrate Chemistry for Food Scientists, St. Paul, MN, American Association of Cereal Chemists.

Wurzberg, O. B. (1986). Cross-linked starches. In O. B. Wurzberg, (Ed.), *Modified Starches: Properties and Uses* (pp. 41-53). Boca Raton: CRC Press.

In: Advances in Cereal and Pseudocereal ... ISBN: 978-1-62618-347-6
Editors: N. Morita, P.V. Hung et al.

Chapter 9

APPLICATION OF CHEMICALLY MODIFIED STARCHES FOR BREADMAKING

Pham Van Hung[1], Megumi Miyazaki[2] and Naofumi Morita[3]

[1]School of Biotechnology, International University, National University in HoChiMinh City, HoChiMinh City, Vietnam

[2]Central Laboratory, Yamazaki Baking Co., Ltd., Chitose, Sumida-ku, Tokyo, Japan

[3]Department of Food Packaging Technology, Toyo College of Food Technology, Mimami-Hanayashiki, Kawanishi, Hyogo, Japan

ABSTRACT

Modified starches have been investigated and developed over a century and have various applications in food, paper and textile industries. Recently, chemically modified starches such as esterified, etherified and/or cross-linked starches have been widely used to enhance the quality of various prepared foods such as snacks, breads and cakes. In breadmaking, starch is a major component and plays an important role in texture and quality of dough and bread. Therefore, modified starches, which are developed to improve undesirable properties of native starches, also affect the properties of dough and quality of breads. The modified starches could be used to substitute up to 20% of wheat flour without compromising the quality of breads. However, vital gluten needs to be blended with the modified starches to improve loaf volume of breads. As a result, a suitable amount of modified starches could be used for breadmaking to improve the functionality and quality of breads. In this chapter, the dough properties and breadmaking qualities of wheat flour substituted with various modified starches such as hydroxypropylated, acetylated and/or cross-linked corn starches are reviewed with the hope that it could provide the bakeries with useful information about a supplemental material for wheat flour in breadmaking.

INTRODUCTION

Nowadays, resistant starches (RS) are widely supplemented into different foods due to their health benefits like dietary fiber with being almost totally fermented in the colon to short-chain fatty acids that help to prevent colorectal cancer, to lower risk of heart disease, and to influence metabolic and inflammatory bowel diseases such as diabetes and diverticulitis (Craig et al., 1998; Topping & Clifton, 2000). In 2001, 500 new products were launched globally with benefit statements regarding their fiber content and almost 50% of them were cereal products - bakery products, breakfast cereals, and energy bars, etc. (Kamp & Jones, 2003).

Some previous studies reported that chemically modified starch also had the characteristics of resistance to enzymatic hydrolysis like resistant starch (Eerlingen et al., 1993; Brown et al., 1995; Woo & Seib, 2002). The modified starches with high resistant starch content are prepared by reaction with various functional groups such as acid anhydrides, phosphoryl chloride, sodium trimetaphosphate, etc. Woo and Seib (2002) found that cross-linked starches prepared from wheat, corn, and potato starches by a moderately rapid reaction in aqueous slurry of the mixture of sodium trimetaphosphate (STMP) and sodium tripolyphosphate (STPP) gave 58-75% RS when assayed with the total dietary fiber method of the Association of Official Analytical Chemists (AOAC).

Although the increase in dietary fiber (or resistant starch) content in food products has certain health benefits, the sensory and textural properties of these foods are adversely affected. In breadmaking, staling is an important factor to evaluate the quality of bread and the retrogradation of starch is believed to be responsible for this phenomenon. On cooling and aging of bread, staling is caused by the re-association of amylose and amylopectin that leach out from starch granules during gelatinization in baking process. The gelation of amylose and crystallization of amylopectin are known to involve in retrogradation of starch and the crystallization of amylopectin is considered to be the main cause of bread staleness. Some studies reported that the substitution of wheat flour with waxy wheat starch retarded the retrogradation of starch in bread during storage (Lee et al., 2001; Morita et al., 2002). However, the retrogradation of starch is also affected by the existence of other linear fractions of starch. The presence of cross-linked structure increases the gelatinization temperature, setback viscosity and decreases the transition enthalpy of gelatinization (Woo & Seib, 2002; Zheng et al., 1999). Paste properties of starch granules might be modified by cross-linking to alter the texture and rheology of the paste. Previous studies reported that the bread substituted with cross-linked starches increased firmness and other characteristic of breadcrumb (Bechtel, 1959; Inagaki & Seib, 1992). However, they did not show a clear relationship between swelling degree of cross-linked starches and the texture of bread. In this chapter, effects of different levels of modification and amylose contents of the cross-linked corn starches on physicochemical and rheological properties of dough and quality of bread are reviewed.

Dough Properties of Modified Starch Substituted Wheat Flours

Wheat protein is about 12% (anhydrous basis) in endosperm of the grain, and possesses the unique property of forming gluten when wetted and mixed with water. Gluten imparts physical properties that differ from those of doughs made from any other cereal grains. When water is added and mixed with wheat flour, the water-soluble proteins together with the water-insoluble proteins hydrate and form gluten, in which starch and other components are embedded (Pomeranz, 1988). Thus, the gluten is, in reality, the skeleton of wheat flour dough, and plays an important role in gas retention, which makes the light, leavened products. Therefore, the substitution of starches for some parts of wheat flour decreases total weight of wheat protein, and results in weaker skeleton of wheat flour dough. To obtain sufficient loaf volume, and good mixing or mechanical resistance of dough, vital gluten is needed as an improver. The amount of vital gluten added is about 8% of the weight of starch substitution (Hung et al., 2004; Miyazaki et al., 2005a; Miyazaki et al., 2005b). This weight is almost equal to the percentage of insoluble wheat protein, glutenin and gliadin in wheat flour (Pomeranz, 1988).

As the amount of starch substitution increases, it is expected that the dough and the bread takes on nature of the substituted starches. In case of using modified starches other than wheat starch derivatives, such as modified corn and tapioca starches, the amount of starch substitution is desirable at less than 20% of wheat flour (Hung et al., 2004; Miyazaki et al., 2005a; Miyazaki et al., 2005b).

At the presence of suitable amount of vital gluten, water absorption of wheat flour dough substituted with modified starches is affected by both nature of starch and types of modification. Several studies reported that doughs made from waxy wheat flour or portion of waxy wheat flour absorbed more water than those made from non-waxy wheat flour (Lee et al., 2001; Morita et al., 2002; Abdel-Aal et al., 2002). Therefore, dough containing 5 - 10% of cross-linked waxy corn starch (CWCS) and vital gluten increases water absorption by 0.2 - 1.2% as compared with 100% of wheat flour (Hung et al., 2004). In contrast, substitution with non-waxy starch derivatives usually decreases water absorption of the dough (Table 9.1). Dough containing 5 - 10% of cross-linked non-waxy corn starch (CNCS) or 20% of several chemically modified tapioca starches with vital gluten decreases water absorption by 1.2 - 2.2%, and 1.5 - 4.0%, respectively, as compared with 100% of wheat flour (Hung et al., 2004; Miyazaki et al., 2005a). Also, substitution with 5 - 10% of cross-linked corn starches without addition of vital gluten decreases the water absorptions of dough (Hung et al., 2005). However, the substitution of hydroxypropylated starches slightly increases water absorption of dough made from the same amount of native starch, or wheat flour alone. The substitution of hydroxypropylated cross-linked potato starch for 10 - 30% of wheat flour increases water absorption as compared with the same substitution of native potato starch (Takasaki & Mineki, 2001). Likewise, the substitution of hydroxypropylated tapioca (Miyazaki et al., 2005a) or wheat starch (Ono et al., 1999) for 20 - 30% of wheat flour also increases water absorption as compared with the same substitution of native starch, or wheat flour alone. Other chemically modified starches, like acetylated starches, generally decrease water absorption of the wheat flour dough at the less than 20% of substitution (Miyazaki et al., 2005a). Starch is present in the dough in the native state and absorbs up to about 46% water during dough preparation (Goesaert et al., 2005). After cross-linking, the specific chemical reacts with two different hydroxyl groups on the same molecule or on different molecules of amylose and/or amylopectin and acts as bridges between molecules. As a result, the granules become compact and absorb less water than the native starch. In contrast, the hydroxypropylated starch absorbs more water and increases water absorption of starch-substituted doughs because hydroxylpropyl groups are hydrophilic in nature.

Table 9.1. Farinogragh data of doughs made from flours substituted with cross-linked cornstarches[a,b,c] (Adapted from Hung et al., 2004)

Sample	Water absorption (%)	Time (min)			Weakness (BU)
		Arrival	Development	Stability	
Control	61.2 ± 0.4	6.5	15.5	14.0	40
5L-CNCS(G)	60.0 ± 0.2	2.0	13.0	17.0	40
10L-CNCS(G)	59.8 ± 0.2	2.0	14.0	18.0	40
15L-CNCS(G)	59.6 ± 0.2	0.5	12.0	17.0	40
5H-CNCS(G)	59.8 ± 0.2	2.0	14.0	18.0	40
10H-CNCS(G)	59.4 ± 0.2	1.5	14.5	19.5	35
15H-CNCS(G)	59.0 ± 0.2	0.5	15.5	21.0	40
5L-CWCS(G)	61.4 ± 0.4	4.0	14.5	15.5	30
10L-CWCS(G)	62.0 ± 0.4	1.5	15.0	18.5	30
15L-CWCS(G)	62.4 ± 0.4	0.5	12.5	16.5	40

[a] Abbreviations: L-CNCS(G), low-swelling cross-linked non-waxy cornstarch containing 8% vital gluten; H-CNCS(G), high-swelling cross-linked non-waxy cornstarch containing 8% vital gluten; L-CWCS (G),low-swelling cross-linked waxy cornstarch containing 8% vital gluten; 5, 10 and 15 mean the amount of substitutions (%).

[b] The value shows the the mean of duplicate data.

[c] BU, Brabender unit.

Speed of water absorption into wheat flour (arrival time on farinogram chart) and development time of dough (peak time on farinogram chart) containing 20% of modified tapioca starches becomes short, but mixing tolerance is the same or slightly increases at the presence of vital gluten (Hung et al., 2004; Miyazaki et al., 2005a).

Substitution of cross-linked corn starch (CLCS) and vital gluten for wheat flour improves the texture of dough i.e., making the dough stronger and more extensible than wheat flour during proofing, and this is more effective by using CLCS with higher swelling degree (H-CLCS), 7.0 - 9.0 (Table 9.2) (Hung et al., 2004). Specific volume of bread baked from the substituted wheat flour with CLCS and vital gluten is almost the same, and especially, the bread substituted with H-CLCS has bigger specific volume than wheat flour-bread or bread containing cornstarch with lower swelling degree (L-CLCS),

2.0-2.8 (Hung et al., 2004). These results mean that the level of cross-linkage of starches affects the cohesive, rubbery texture of the doughs during molding, proofing, and baking. However, the role of native and modified starches in the dough is still ambiguous.

Table 9.2. Extensional properties of doughs made from flours substituted with cross-linked cornstarches[a,b,c] (Adapted from Hung et al., 2004)

Sample	R (BU)			E (cm)			R/E		
	45 min	90 min	135 min	45 min	90 min	135 min	45 min	90 min	135 min
Control	373ab	675b	767cd	13.2a	12.1a	10.1a	28.3cd	55.8d	75.7f
5L-CNCS(G)	355a	673b	747bcd	18.2d	14.3bc	14.4def	19.5a	47.2bc	51.7bc
10L-CNCS(G)	407b	662b	805d	16.5c	15.7cd	14.2def	24.6bc	42.1ab	56.8cd
15L-CNCS(G)	535f	765c	870e	15b	14.7cd	14def	35.6ef	52.2cd	62.1de
5H-CNCS(G)	523f	752c	915e	16.5c	15.9d	13.1cd	31.7de	47.3bc	69.7ef
10H-CNCS(G)	465de	770c	908e	13.7a	12.9ab	11.8bc	33.9e	59.5de	77.2f
15H-CNCS(G)	507ef	815c	885e	12.8a	12.3a	11.3ab	39.5f	66.3e	78.1f
5L-CWCS(G)	348a	568a	643a	15.6bc	15.3cd	15.5f	22.3ab	37.1a	41.4a
10L-CWCS(G)	413bc	583a	690ab	15.7bc	14.2bc	14.8ef	26.3bc	41.1ab	46.6ab
15L-CWCS(G)	438cd	675b	730bc	15.2b	14.8cd	13.5de	28.8cd	45.6abc	54.2bcd

Roles of the Modified Starches in Texture and Staling of Bread

Bread crumbs prepared from the flour substituted with native, hydroxypropylated and acetylated tapioca starches have a tacky texture, while breads with phosphorylated cross-linked tapioca starch have a dry feel (Miyazaki et al., 2005a; Miyazaki et al., 2005b). The flour substituted with the former starches shows higher peak viscosity and large breakdown while that with the later shows lower peak viscosity and smaller breakdown than wheat flour on the amylogram (Miyazaki et al., 2005a). These results mean that viscous texture of gelatinized starch affects texture of bread. By using modified starches, bakers can control texture of products, and develop unique breads which have different texture as compared to wheat flour-bread.

The decline of bread quality such as a decrease in flavor impact or a drier mouth-feel during storage is called "staling", or bread staling (Stauffer, 2000). Factors affecting bread staling have been extensively investigated, and rearrangement in the starch fraction is a significant base of staling (Hug-Iten et al., 2003).

Some workers studied quality or staling of bread by substituting chemically modified starch for wheat starch (Inagaki & Seib, 1992; Toufeili et al., 1999) or wheat flour (Takasaki & Mineki, 2001; Hung et al., 2004; Hung et al., 2005; Miyazaki et al., 2005b). Inagaki and Seib (1992) and Toufeili et al. (1999) reported that bread with cross-linked waxy barley starch with 50 - 500 ppm phosphorous oxychloride staled easily and showed a higher enthalpy change in melting of crystalline region of starch. Breads prepared from substituted flour with phosphorylated cross-linked tapioca starch have harder crumb at the baked day, and stale more easily than the bread substituted with tapioca starch and wheat flour (Miyazaki et al., 2005b). Although breads substituted with 5 - 10% of CLCS and vital gluten for wheat flour have bigger loaf volume and soft crumb, the firmness of breads increases with increasing amounts of CLCS (Figure 9.1) (Hung et al., 2004). Breads substituted with CWCS with or without addition of vital gluten for wheat flour has soft crumb during storage, and the firmness of bread crumb is the same in spite of increasing amount of CWCS (Hung et al., 2004; Hung et al., 2005). This means that some waxy starches possess anti-staling ability, although substitution of cross-linked non-waxy starches for wheat flour generally fastens bread staling. Because cross-linked starch granules are stable during baking, the bread containing cross-linked starch remains the large starch granules (around 20 μm diameter) in breadcrumb. The remaining of granules in crumb and their entanglement with gluten network is considered to be one of the factors that affect hard texture of bread (Miyazaki et al., 2005b). In addition, the bread containing cross-linked starches shows larger staling endothermic enthalpy during storage than that from wheat flour, which is affected by the faster staling (Toufeili et al., 1999; Miyazaki et al., 2005b). Bread containing 30% of cross-linked hydroxypropylated potato starch shows softer crumb than that containing native potato starch (Takasaki & Mineki, 2001). Firmness of bread containing 20% of hydroxypropylated tapioca starch or acetylated tapioca starch and vital gluten is less than the bread containing the same amount of tapioca starch or wheat flour alone. The retrogradation endothermic enthalpy of stale bread baked from substituted flour with hydroxypropylated tapioca starch is less than the bread baked from wheat flour alone (Miyazaki et al., 2005a). These results indicate that firming of bread with various chemically modified tapioca starches is associated with retrogradation of amylopectin.

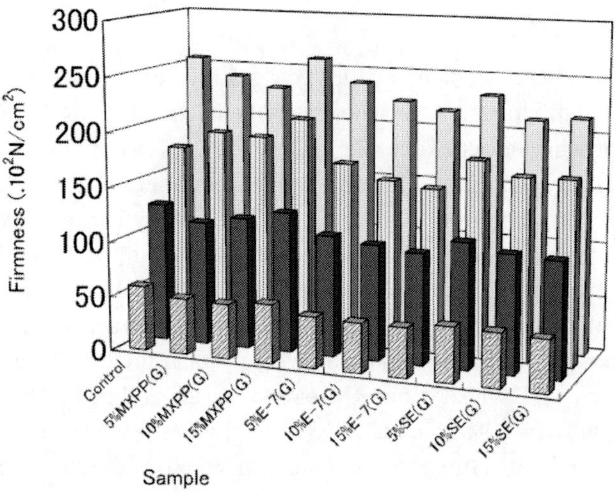

▨, 0 day; ■, 1 day; □, 3 days; ▤, 5 days.

Figure 9.1. Firmness of breads baked with various amounts of cross-linked cornstarches and vital gluten which substituted for wheat flour. Abbreviations are the same in Table 9.1 (Adapted from Hung et al., 2004).

Thus, many factors of starch affect breadmaking properties, but gelatinization and retrogradation characteristics are considered to be quite important for bread quality. Gelatinization properties affect the texture of bread. Among various modified starches, hydroxypropylated starches are the most effective to retard staling. DSC data shows that doughs containing 20% of hydroxypropylated tapioca starch and vital gluten have the same gelatinization temperature and endothermic enthalpy as wheat flour dough (Miyazaki et al., 2005b). This would be one of the key factors that make hydroxypropylated tapioca starch suitable for breadmaking. As ether or ester modification to starches lowers gelatinization temperature, and changes endothermic enthalpy, they are thought to be good strategies for various non-wheat starches to retard staling in breadmaking. Although large amount of substitution with cross-linked non waxy starches for wheat flour fastens bread staling, cross-linking is an essential method to control stable viscosities of starch gel. Takasaki & Mineki et al. (2001) reported that cross-linked hydroxypropylated potato starch has softer crumb than that containing native potato starch. Therefore, treatment with both hydoxypropylation and cross-linking, or acetylation and cross-linking might be more suitable for breadmaking. However, it is necessary to further study on bread quality of modified starch-substituted flour with different degree or ratio of modification of ester, ether and cross-linking starches, or different starch materials to make the potentiality of modified starches in breadmaking expand.

CONCLUSION

Chemically modified starches such as hydroxypropylated, acetylated and/or crosslinked starches could be used to substitute for wheat flour in breadmaking. The amount of starch-substitution is desirable at less than 20% of wheat flour, and vital gluten should be added at approximately 8% of the weight of substituted starch to obtain sufficient loaf volume, mixing or mechanical resistance. The bread baked with the substitution of 5% or 10% of L-CNCS or H-CNCS and vital gluten for wheat flour had bigger loaf volume and softer breadcrumb than did the control. The substitution of higher-swelling-degree CLCS made the stronger dough, bigger loaf volume and softer breadcrumbs than that of the lower-swelling-degree CLCS. In addition, the substitution of cross-linked waxy cornstarch for wheat flour improved the softness of breadcrumb during storage. These results indicate that modified starches with superior functionality will be commercially valuable, and widely used as bread improvers in the future.

REFERENCES

Abdel-Aal, E.-S. M., Hucl, P., Chibbar, R. N., Han, H. L., & Demeke, T. (2002). Physicochemical and structural characteristics of flours and starches from waxy and nonwaxy wheats. *Cereal Chemistry*, *79*, 458-464.

Bechtel, W. G. (1959). Staling studies of bread made with flour fractions. V. Effects of a heat-stable amylase and a cross-linked starch. *Cereal Chemistry*, *36*, 368.

Brown, I. L., McNaught, K. J., & Moloney, E. (1995). New direction in starch technology and nutrition. *Food Australia*, *47*, 272-275.

Craig, S. A. S., Holden, J. F., Troup, J. P., Auerbach, M. H., & Frier, H. I. (1998). Polydextrose as soluble fiber: Physiological and analytical aspects. *Cereal Food World*, *43*, 370-376.

Eerlingen, R. C., Crombez, M., & Delcour, J. A. (1993). Enzyme resistant starch. I. Quantitative and qualitative influence of incubation time and temperature of autoclaved starch on resistant starch formation. *Cereal Chemistry*, *70*, 339-344.

Goesaert, H., Brijs, K., Veraverbeke, W. S., Courtin, C. M., Gebruers, K., & Delcour, J. A. (2005). Wheat flour constituents: How they impact bread quality, and how to impact their functionality. *Trends in Food Science and Technology*, *16*, 12-30.

Hug-Iten, S., Escher, F., & Conde-Petit, B. (2003). Staling of bread: Role of amylose and amylopectin and influence of starch-degrading enzymes. *Cereal Chemistry*, *80*, 654-661.

Hung, P. V., & Morita, N. (2004). Dough properties and bread quality of flours supplemented with cross-linked cornstarches. *Food Research International*, *37*, 461-467.

Hung, P. V., & Morita, N. (2005). Thermal and rheological properties of dough and bread as affected by various cross-linked cornstarch substitutions. *Starch/staerke*, 57, 540-546.

Inagaki, T., & Seib, P. A. (1992). Firming of Bread Crumb with Cross-Linked Waxy Barley Starch Substituted for Wheat Starch. *Cereal Chemistry*, *69*, 321-325.

Kamp, J. W. V. D., & Jones, J. M. (2003). A focus on dietary fiber and whole grains. *Cereal Food World*, *48*, 41.

Lee, M. R., Swanson, B. G., & Baik, B. K. (2001). Influence of amylose content on properties of wheat starch and breadmaking qualities of starch and gluten blends. *Cereal Chemistry*, *78*, 701-706.

Miyazaki, M., Maeda, T., & Morita N. (2005a). Gelatinization properties and bread quality of flours substituted with hydroxypropylated, acetylated and phosphorylated cross-linked tapioca starches. *Journal of Applied Glycoscience*, *52*, 345-350.

Miyazaki, M., Maeda, T., & Morita N. (2005b). Starch retrogradation and firming of bread substituted with hydroxypropylated, acetylated and phosphorylated cross-linked tapioca starches for wheat flour. *Cereal Chemistry*, *82*(6), 639-644.

Morita, N., Maeda, T., Miyazaki, M., Yamamori, M., Miura, H., & Ohtsuka, I. (2002). Effects of substitution of waxy-wheat flour for common flour on dough and baking properties. *Food Science and Technology Research*, *8*, 119-124.

Ono, K., Seib, P. A., & Takahashi, H. (1999). Effects to rheological and tasting properties of noodle with hydroxypropyl wheat starch. *Nippon Kasei Gakkaishi*, *49*, 985-992 (in Japanese).

Pomeranz, Y. (1988). Composition and functionality of wheat flour components. In Y. Pomerantz (Ed.), *Wheat: Chemistry and Technology*, Vol.2, (p.333). St. Paul, Minnesota: American Association of Cereal Chemists.

Stauffer, C. E. (2000). Emulsifiers as antistaling agents. *Cereal Foods World*, *45*, 106-110.

Takasaki, S., & Mineki, M. (2001). Effects of potato starch and modified potato starch on breadmaking properties. *Journal of Cookery Science of Japan*, *34*, 53-61 (in Japanese).

Topping, D. L., & Clifton, P. M. (2001). Short-chain fatty acids and human colonic function: Roles of resistant starch and nonstarch polysaccharides. *Physiological Reviews*, *81*, 1031-1064.

Toufeili, I., Habbal, Y., Shadarevian, S., & Olabi, A. (1999). Substitution of wheat starch with non-wheat starches and cross-linked waxy barley starch affects sensory properties and staling of arabic bread. *Journal of the Science of Food and Agriculture*, *79*, 1855-1860.

Woo, K. S., & Seib, P. A. (2002). Cross-linked resistant starch: Preparation and properties. *Cereal Chemistry*, *79*, 819-825.

Zheng, G. H., Han, H. L., & Bhatty, R. S. (1999). Functional properties of cross-linked and hydroxypropylated waxy hull-less barley starches. *Cereal Chemistry*, *76*, 182-188.

INDEX

α

#

A

B

N

O

P

Q

R

S

T

U

V

W

X

Y